THEORIES of the EVOLUTION of KNOWING

The T. C. Schneirla Conference Series Volume 4

Edited by

Gary Greenberg
The Wichita State University

Ethel Tobach
The American Museum of Natural History

LEA LAWRENCE ERLBAUM ASSOCIATES, PUBLISHERS
Hillsdale, New Jersey Hove and London

Lawrence Erlbaum Associates, Inc., Publishers
365 Broadway
Hillsdale, New Jersey 07642

Library of Congress Cataloging-in-Publication Data

T.C. Schneirla Conference (4th : 1987 : Wichita State University)
Theories of the evolution of knowing : proceedings of the fourth T.C. Schneirla Conference / edited by Gary Greenberg, Ethel Tobach.
p. cm.
Proceedings of the fourth T.C. Schneirla Conference held at Wichita State University in November 1987.
Includes bibliographical references and index.
ISBN 0-8058-0755-1
1. Knowledge, Theory of--Congresses. 2. Evolution--Philosophy--Congresses. I. Greenberg, Gary. II. Tobach, Ethel, 1921- .
III. Title.
BD177.T2 1987
155.7--dc20 90-43391
CIP

Printed in the United States of America
10 9 8 7 6 5 4 3 2 1

Contents

DISCUSSION

PREFACE

The fourth conference honoring T. C. Schneirla's contributions to psychology was held once again at The Wichita State University in November 1987. Our theme focused on evolutionary epistemology (i.e., the evolution of knowledge and of knowing). We were fortunate in having Donald Campbell, one of the important contributors to this area of study, to present the opening paper and to thus set the stage for an interesting series of sessions and theoretical discussions. The papers presented depict current thought about the topic, as well as current criticism of it. The panelists represented diverse areas of study: comparative psychology, developmental psychology, physiological psychology, cultural psychology, and philosophy. As has been the tradition in the conference series, the final session allowed all the panelists, discussants, and organizers to interact as a group during a 3-hour discussion. This provided for the airing of many theoretical points and the clarification of many raised during the paper presentations. Authors utilized this discussion in the preparation of their final drafts.

We are indebted to the administration of The Wichita State University for its continued funding of the conference series.

Gary Greenberg
Ethel Tobach

1 Levels of Organization, Downward Causation, and the Selection-Theory Approach to Evolutionary Epistemology

Donald T. Campbell
Lehigh University

My task in this chapter is to relate evolutionary epistemology to Schneirla's several "levels of integration" themes. It is a task that I welcome because it is needed in revising and clarifying my own *selection-theory approach to the theory of knowledge*.

A 600-item bibliography now exists for evolutionary epistemology (Campbell, Heyes, & Callebaut, 1987), and I have dabbled in the area for over 30 years. From this too-vast enterprise, I have chosen themes that I share with T. C. Schneirla. These include: (a) integrative levels (b) an emphasis on physicalism (materialism) and physical mechanisms, a physicalism that nonetheless avoids atomistic reductionism, doing this through recognizing emergent levels in organic and cultural developmental history; (c) a focus on integrative levels that avoids idealistic vitalism and teleological causation as emphatically as it does atomistic physicalistic reductionism; (d) an effort to reconcile the autonomy of each level of organization with the fact that, in biological systems at least, wholes are a determinant of the nature of parts, as well as parts being the determinants of the nature of the wholes.

Figure 1.1 is an effort to depict several "levels" concepts that are relevant to Schneirla and the related literature. Dimension I, "Encompassing Levels," is a part of the background that Schneirla sought to modify rather than reject. He and Topoff (1984, pp. 241–244) employ it briefly as a starting point. They find the concept of "the superorganism" an inadequate theoretical basis for understanding social insect colony organization but do not reject the reality of the level of organization to which it points. Dimension 1 (as illustrated, ending with the social insect colony) is central to the following section on "Downward Causation." The nonbiological ways in which the Social Organization level is exem-

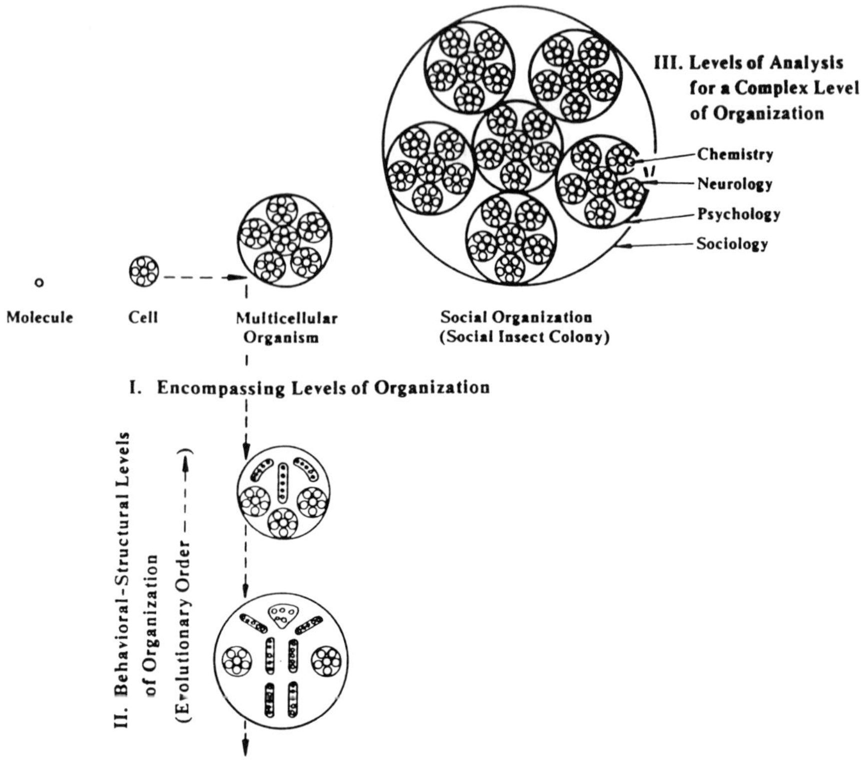

FIG. 1.1. Three dimensions of "Levels of Integration."

plified in humans are too complex for this chapter (but see Campbell, 1983). Users of this series of levels often include "species" as a higher level, instead of "Social Organization." Not yet being convinced that species are a level of *organization*, I demur on this.

My draftsmanship for this dimension has been lazy and oversimplified in a very significant way. The encompassed level is portrayed as if made up of identical units. On the contrary, it seems to be a general rule that where the next higher level is a real level of organization, the next lower level is composed of complementary diversity. Thus the biochemical molecules are made up of diverse, mutually fitting atoms; the cell is made up of diverse molecules and molecule types (e.g., diverse proteins, ribonucleic acids); the multicellular organism is made up of nerve cells, bone cells, muscle cells, etc.; and the social insect colony is made up of fertile females and males as well as diverse sterile castes.

More extensively used in the Schneirlaian literature are the other dimensions. Dimension II is Behavioral-Structural Levels of Organization (primarily among the multicellular organisms, but including their unicellular ancestors). This dimension in Fig. 1.1 stands for several parallel subdimensions. Most biostructural are uses such as Aronson's (1981) four levels of complexity in the chordates up to the reptiles, and Jensen's (this volume) levels in the many evolutionary prestages of human beings. Most used of Dimension II are the levels of *social* behavioral organization (Tobach, 1981; Tobach & Schneirla, 1968; Greenberg, 1988): (a) taxis; (b) biotaxis; (c) biosocial; (d) psychotaxis; and (e) psychosocial. My "levels" in the purported "nested hierarchy of vicarious knowledge processes" (Campbell, 1974b, and following) are, overall, probably best assimilated to this dimension. But my focus is on "knowledge" of the "external" world, and this results in my treating conspecific sociality as "internal." A reformulated set of Dimension II levels for my evolutionary epistemology might begin with: (a) subtaxis "tumbling"; (b) taxis; (c) distance receptors and object permanence; and (d) socially vicarious.

Dimension III, Levels of Analysis, characterizes much of Schneirlaian usage, with particular emphasis on the psychological level. Note that Dimensions I and II could be the edges of a square matrix that could be filled out completely; Dimension II extension is available for each level of Dimension I. For any complex level of organization, a Dimension III Levels of Analysis can be employed, as illustrated in Fig. 1.1 for Social Organization. Dimension III, considered jointly with Dimensions I and II, produces a 3-dimensional solid with a triangular corner unoccupied (e.g., single-celled organisms lack the psychological and sociological levels of analysis).

DOWNWARD CAUSATION

These are exciting times for evolutionary theory. On the one hand, the theory of inorganic physics is being expanded to include more lifelike properties (Prigogine, 1980). On the other hand, and seemingly in conflict, the perceived gap between inorganic and organic forms is being increased through emphases on structural factors in organic evolution. (This literature is reviewed by Wake & Larson 1987, and J. H. Campbell, 1986). I (1987a, pp. 169–175) reviewed literature on "internal selection" and, in so doing, included the theory of punctuated equilibrium. "Downward Causation" belongs among those physicalist theories that stress the organic–inorganic gap.

Both trends reject "bean bag genetics," the notion that single genes have adaptive values that can be stated independently of the context of other genes, a view that implicitly assumes that evolution could have assembled the genome in any order, neglecting any role for organic form and thus implicitly positing a "bean bag organism" as well. We can now see that these aspects of evolutionary theory are at least in part due to a misguided theory of science that made atomistic reductionism the only acceptable mode of explanation (Depew &

Weber, 1986). Whereas biologists and physicists are beginning to reject this outmoded requirement, mainstream philosophy of science has not yet gotten there (e.g., Klee, 1984). The case of Circk (1966) illustrates that not all (perhaps not even most) biologists have made the shift.

Most of us in this volume probably start with the physicalist world view (or origin myth): "In the beginning there was the inorganic world. Life (and later, knowing life) emerged in this inorganic world without contradicting any of the laws of inorganic physics or chemistry." Many of us probably also accept the hypothesis of several ontologically "real" levels of organization (e.g., subatomic particles, atoms, molecules, cells, multicellular organisms, and social organizations). For most of us teleological causation is rejected, but purpose-like achievements in biological organisms are recognized. In accordance with many biologists the term *telenomy* (e.g., Monod, 1971; Wimsatt, 1972) can be used to designate our position.

Within these assumptions, I propose four principles for the relationship between organic and inorganic laws along Dimension I of Figure 1.1 (reworded slightly from Campbell, 1974a, p. 180):

1. All processes at the higher levels are restrained by and act in conformity to the laws of lower levels, including the levels of subatomic physics (a "reductionist" constraint).
2. The telenomic achievements at higher levels require for their implementation specific lower level mechanisms and processes. Explanation is not complete until these micromechanisms have been specified (a "reductionist" constraint).
3. (The emergentist principle) Biological evolution, in its meandering exploration of segments of the universe, encounters laws, operating as selective systems, which are not described by the present laws of physics and inorganic chemistry, and which will not be described by the future substitutes for the present approximations.
4. (Downward causation) Where natural selection operates through life and death at a higher level of organization, the laws of the higher level selective system determine in part the distribution of lower level events and substances. Description of an intermediate-level phenomenon is not completed by describing its possibility and implementation in lower level terms. Its presence, prevalence, or distribution (all needed for the complete explanation of biological phenomena) will often require reference to laws at a higher level of organization as well. Paraphrasing Point 1, for biology, all processes at the lower levels of a hierarchy are restrained by, and act in conformity to, the laws of the higher levels.

Ecological Niche and Selective Aggregate

The disagreement with those who deny points 3 and 4 will have to be settled by judgments of plausibility in the arena of contingent, descriptive statements, and not logical proof. The microdeterminists would probably not claim that science is *already* able to understand or explain (to say nothing of predict) the shape of a soldier ant's jaw, for example, using only the principles of subatomic physics (or even adding those of inorganic chemistry). Nor am I claiming a logical proof of the impossibility of so doing. What I assert is that the plausibility arguments are overwhelmingly in favor of substantial macrodetermination. I argue that we have no grounds for anticipating that the microparticulate laws, even when perfected, will eliminate the need for macrodetermination involving laws unique to the higher levels of biological and social organization.

Crystal Formation. Crystal formation and related processes can be used to characterize instances in which micropartical attributes explain macrostructures. In solutions of salt in water near saturation, at some intermediate temperature salt crystals will begin to form. Relative to the molecules of NaCl and the atoms of Na and Cl, these are macrostructures. Here, the attributes of the subatomic particles plus the background conditions predict the quasi-stable atoms, the attributes of which predict the quasi-stable molecules, the attributes of which predict the quasi-stable salt crystals. The background conditions, water molecules, temperatures, saturation, etc., are perhaps all understandable in terms of the science of microparticles. Many issues of distribution (e.g., "Why this crystal shape rather than one of the few others NaCl permits?" "Why NaCl rather than other compounds?") may be meaningfully addressed in terms of theory developed for microparticles.

In addition to crystal formation, there are other macroprocesses and distributions understandable in terms of the laws of microparticles, such as the occurrence of organic chemical molecules in interstellar space or on earth in an origin-of-life scenario. Prigogine's (1980) nonequilibrium thermodynamics and dissipative structures probably should also be included. These theories greatly increase the explanatory capacity of microdeterminism. From the point of view being advocated here, these macrostructures provide the raw material of variations on which biological natural selection can operate. They are, however, inadequate to predict or explain which variant organic forms will be most abundant.

This is, of course, a matter of degree. Viruses can often be crystalized. Microscopic photographs of the skeletons of diatoms make it seem likely that the forms of the skeletons might be to a considerable extent explained by crystal-formation-like considerations, even though we also know that selection for whole-organism survival has been involved. Such microparticulate explanations go much less of the way toward explaining the shape of a giraffe, although its bilateral symmetry may have microprocess roots, rather than roots in selection for whole-animal functionality.

The Quasi-entropic Effect of Mutations. It may help highlight the relation of distributional facts to macrodetermination of "downward causation" if we consider dysfunctional mutations. In the classic 50 years of *drosophila* research, many more harmful mutations than beneficial ones were discovered—on a ratio in excess of 100 to 1. The initial interpretation of these findings made the organism seem extremely vulnerable to one-locus mutations. This picture has been greatly modified by modern discoveries of the ubiquity of neutral alleles, the numerous repetitions of the most fundamental genetic instructions in the chromosonal material, and DNA repair mechanisms. Nonetheless, the evidence from Morgan's laboratories remains, and a general tendency for mutations to be deleterious, however slightly, seems undeniable. This produces a general tendency for intricate, functional order to drift toward dysfunctionality in the absence of continual selection for function. The classic illustration is of cave fish becoming blind when vision loses its survival value, when mutations interfering with effective vision cease to be eliminated. (Whereas the selective disadvantages of eyes as a site for infection and as metabolic cost are relevant, these do not explain those instances in which the gross anatomy of the visual system is still retained.)

The point of this tangent is that, in terms of all the laws and attributes of the physical microparticles (both as presently known and in any anticipatable asymptotic perfection), *the dysfunctional mutants are at least as lawful outcomes of microparticulate laws as are the normal genes. The laws of physics do not choose between them.* The distributional fact that the normal gene with the functional expression is more prevalent cannot be explained by particle physics. This prevalence is due to selection at the whole-animal (multiple cellular, multiple molecule, multiple atomic), levels, a selection that has affected the distribution of the particular variants of all lower units. This is the "downward causation" via selection. It is a local instance of macrolevel determination of the *distribution* of microlevel events and entities.

Level-specific Functional Laws as Constraints. Consider the anatomy of the jaws of a worker ant. The hinge surfaces and the muscle attachments agree with Archimedes' laws of levers, that is, with macromechanics. They are optimally designed to apply the maximum force at a useful distance from the hinge. A modern engineer could make little if any improvement on their design for the uses of gnawing wood, picking up seeds, etc., given the structural materials at hand. This is a kind of conformity to physics, but a different kind than is involved in the molecular, atomic, subatomic strong and weak coupling processes, etc., underlying the formation of the particular proteins of the muscle and shell of which the system is constructed. The laws of levers are one part of the complex selective system operating at the level of whole organisms. Selection at that level has optimized viability and has thus optimized the form of parts of organisms, for the worker ant and for their solitary ancestors. We need the laws of levers, *and organism-level selection*, to explain the particular distribution of proteins found in the jaw and *hence* the DNA templates guiding their production. (The occasional nonfunctional mutant forms of jaws conform just as loyally to the laws of levers and biochemistry as do the more frequent functional forms.) Even if Archimedes' laws of levers can someday be derived from quark theory (which seems unlikely), their partial reflection in an ant's jaws are not to be explained by microdetermination, but rather by macrolevel selection.

If we now consider the jaw of a soldier ant, a still more striking case of emergence and downward causation is encountered. In many of the highly dimorphic or polymorphic species, the soldier's jaws are so specialized for piercing other organisms (huge multipronged antler-pincers) that the soldier cannot feed itself and has to be fed by workers. The soldier's jaws and the distribution of protein therein (and the particular ribonucleic acid chains that provide the templates for the proteins) require for their explanation certain laws of sociology centering around division-of-labor social organization. The syndrome of division of labor, storable nonspoiling foodstuff such as honey or seeds, apartment house living, and professional soldiers who do no food gathering has been repeatedly independently discovered many times among the social insects (Campbell, 1965).

My challenge to microdeterminists is based on the patent implausibility of expecting extensions of *the laws of microparticulate physics* (keeping biological knowledge out of the background conditions employed) helping us understand the difference between the soldier and worker jaws and the relationship of this difference to mutual feeding and defense.

The social insects also provide a "downward causation" example in the relation of sociology and psychology. A full study of the psychology of individual members of an ant colony (their stimulus–response repertoires, reactions to pheromones, and the totality of their behavioral dispositions) might make possible the prediction of the overall sociology of the nest. But if we then inferred that these behavioral dispositions first evolved independently in solitary ancestors who then discovered each other in some analogue to the social contract, we would be wrong. Insofar as differing from those of their solitary and semisocial ancestors, these dispositions, and the concommitant interlocking roles, are in considerable part due to selection at the level of whole-nest survival. Selection at this level usually is not undermined by genetic competition among the cooperators, inasmuch as these cooperators are almost completely sterile (see Campbell, 1983, for a more detailed exposition). Thus substantial aspects of the individual psychology of the ants are due to evolutionary optimizing at the level of functioning social organizations, the nests or colonies.

Schneirla frequently emphasized the emergence of social coordination from the interaction of the individual ant with the momentary environmental complex, including the other ants, the larva, etc. Sometimes this sounds like the social-contract theory just parodied, in which the psychological level (plus environmental contingencies) generates the social coordination, with no attention as to why the psychological dispositions were of such a nature (considerations that would seem to me to require downward causation via colony selection). Probably, however, Schneirla would not have rejected the analysis I have summarized. He was certainly rejecting the clock-work model of insect behavior and accentuating the intelligence and purposiveness of individual insect behavior, the extemporaneous solution of local obstacles, etc. He probably would not have denied that the subgoals and goals of this intelligent extemporizing had been selected, in considerable part, by colony-level selection, nor that this selection accounted for the similarity of the emergent social order found among nests of the same species, and for the different social orders found in other species.

Technically, in the terms of population genetics, it is *not* group selection that is involved, which would be a matter of group selection of queens. Two quotations from Boorman and Levitt (1980) make this clear. ''Mating does not normally take place between reproductives produced by the same social insect colony. Insect colonies are, therefore, not reproductively closed populations and accordingly cannot be treated as demes for the purposes of group selection. In turn, this means that group selection is largely ruled out as an explanation of most cases of insect sociality'' (pp. 13–14). However, Boorman and Levitt do not mean to rule out selection by cooperating social unit. In fact they also say ''If selection is mostly at the colony level, workers can be altruistic to the remainder of the colony'' (p. 41). Although they do not explicitly mention that *net* selection at the colony level can best take place when selection at the individual level (workers and soldiers) has been eliminated by caste sterility, it is probable that they would agree. The evolutionary history of the ants starts with postponed fertility of adult daughter nest helpers. When such helpers become a requisite for nest survival, the inclusive fitness interests of a sterile worker come to include keeping fellow workers sterile. At the early stage, fertility is the only division of labor, the protoqueen and her daughters share all the other tasks, such as foraging, defense, brood care. Mechanisms for extending and maintaining worker sterility make possible the subsequent evolution of soldier and other sterile caste specializations (see Campbell, 1983, for more details).

Ecological Niche and Selective Aggregate. My initial presentation (Campbell, 1974a) was weakened by careless usage of the phrase ''selective system.'' ''Quasi-stable ecological hodgepodge'' would have been a better phrase. The ''nature'' in natural selection need not and should not be assumed to be systematic in any sense other than being a somewhat stable composite, persisting over numerous reproductive generations. There may be systematic components to the aggregate, as in diurnal and annual cycles and in the behavior of other animals and plants, but there is usually no justification for regrading the composite of systematic and singular features of the ecology as forming a ''system.'' Most so-called ecosystems are instead ecoaggregates. Positing system at this level should be avoided unless the node of selection for this level can be specified (as it can be for some two-species symbioses, but rarely for more than two). No selective ''system'' is required for the examples of macrodetermination I have brought forward.

The ''higher level laws'' ''encountered'' by higher levels of organization are thus *not* to be found in the ''system'' or ''organization'' of the *physical* environment (including other organisms and their behavior in that physical environment). The selective aggregate in this sense may well have been shared by the social ant and its solitary ancestors.

The "laws" operating as "selectors" for the shape of the jaws of worker and soldier ants need conceptual clarification by philosophers of science and evolutionary biologists. I have failed to provide this. The phrase "ecological niche" would provide a better usage than either "selective system" or "selective aggregate," because it connotes a way of earning a living and is somehow closer to the animal's form. However, multicellular organisms will usually have moved into an ecological niche differing from their unicellular ancestors. Multicellular organisms with skeletons or exoskeletons are in a different niche than their flaccid cousins, a niche that includes laws of levers of rigid bodies. Social insects with sterile castes are in a niche substantially different from their solitary predecessors. In some important sense, *a species "constructs" its own ecological niche, rather than passively stumbling into one already "assembled"* by the physical environment and other species (Lewontin, 1982). However, the striking instances of convergent evolution, as in the worker–soldier caste system of many social insects, show that, in this "construction," restraints provided by the inorganic world and "the laws of sociology" also are operative and can produce independent "discoveries." "Construction" and "invention" are prerequisites for discovery, although not sufficient, in that most constructions fail to be discoveries.

A Node of Selection for Every Level of Organization. In an extension of the preceding illustrations, I propose a broad generalization: *Form-structure-order-organization at every level is both achieved and maintained by a node of selection specific to that level.* Neo-Darwinian evolutionary theory has focused on *external selection*, selection to fit the opportunity structure of the external world. A node of such selection clearly operates at the whole-multicellular-organism level for most animals, where death-before reproduction can operate as a selective force. In our current amplification of evolutionary theory, we see this node augmented, amplified, and indeed sharpened by a corresponding whole-multicellular-organism *internal selection* for coherence. (This is more fully discussed in Campbell, 1987a.) For ants and other social insects, there is an *external* node-of-selection at the level of the ant nest. Whole nests survive or die, propogate, or fail to propogate. The workers and soldiers are unable to transfer their participation to another nest nor set up nests of their own. This nest-level (cooperating social unit level) is also augmented and sharpened by internal selection (including those mechanisms that preclude worker transfer to other nests). (For the inorganic levels of organization "internal" nodes of selection can also be classified, however crudely, as exemplified in crystal formation in which the already accumulated crystal acts as a template in the selective retention of other molecules of the same substance.)

Still more conjecturally, I want to see *every* level of organization within the organism as being achieved in development and kept in order by a node of selection. Thus for the multicellular organism, I expect that we will eventually find selective system nodes at the level of DNA molecule (illustrated by DNA "repair" processes), codon, chromosome, centrosome, cell, organ, etc. These nodes of selection for the most part will *not* be *external* in any direct way. Whereas *internal* in this sense, many will differ from crystal formation in that the selector will be an internalized *vicarious* representative of an external selector.

Here is a concrete example (Campbell, 1987a) contrasting *external* and *vicarious-internal* selection:

> In animals like the salamander, capable of regeneration, the loss of a limb removes this template restraint, and the limb regrows. One can imagine this growth being stopped by an *external* selector, the leg growing down 'until it hits the ground.' This is not how it works, however. Instead, the growth processes are restrained, selected, inhibited, or turned off by an internal vicarious selector when a certain length is reached. The vicarious selector represents already evolved wisdom about optimal leg lengths for locomotion and survival. With regard to length, it contains what can be called a 'reference signal' in control system language (Powers, 1973), or a 'homeostat setting' in Ashby's (1952) terminology, analogous to a thermostat setting. This homeostat setting is continually subject to mutations and natural selection, involving the life, death, and differential procreation of whole salamanders, and will be reset at a different value if it becomes seriously non-optimal. Borrowing further metaphors from control system engineering and statistics, one could speak of the vicarious control system for leg length as being 'nested' within the whole-organism-reproduction control system of natural selection. (p. 186)

EVOLUTIONARY EPISTEMOLOGY

As Danilov and Togel (this volume) make clear, there are two major forms of evolutionary epistemology (see also Bradie, 1986; Vollmer, 1987). One is limited to biological evolution, the other often deprecated as merely "analogical." It is the biologizing of knowing (and especially of social behavior) that is the proper target of Lerner's (this volume) stunning chapter. If only to partially deflect that critique from myself, I must stress that my entry into this area was from the "analogical" side. Although later I found dozens of anticipators, it was Ashby (1952) and Pringle (1951) who called my attention to the parallel between natural selection and trial-and-error learning, thus introducing me to the abstract model I now call *selection theory* (following Baldwin, 1909). My first three papers in this project were entitled "Adaptive Behavior from Random Response" (1956a), "Perception as Substitute Trial and Error" (1956b), and "Blind Variation and Selective Retention in Creative Thought as in Other Knowledge Processes" (1960). (This needs stressing if only because the fragment of my evolutionary epistemology that I present here is entirely biological.)

Our conference discussion showed that an explanation is needed as to how both the biological and analogical version are "epistemological" in ways relevant to modern philosopher, whose standard conceptual analysis concludes that *"Knowledge is justified, true belief."* Except for the few modern skeptics who conclude that we have no "knowledge," Anglo–American philosophers have by and large joined G. E. Moore in begging the question of induction, starting the problem of knowledge from the assumption that we have such (rather than starting from total doubt), and accepting for the required "justification" the utilization of usually reliable processes, even though these may sometimes lead to error (Kornblith, 1985). Visual perception of nearby ordinary objects is typically a "justifying" source of belief. As some within this tradition have noted (for references, see Campbell, 1987b; Campbell & Paller, 1989), accepting the theory of biological evolution augments this "justification," making it plausible that, in general and for typical uses, our eyes are trustworthy. This orientation (second only to a species-history explanation of Kant's a priori categories of perception and intuition) is a major theme in biological evolutionary epistemology.

Would *any* theory of evolution provide such "justification," even a creationist one? For most users, the quest for epistemological justification is implicitly or explicitly pushed back behind visual perception into the process of biological evolution, requiring a theory making it epistemologically plausible that biological evolution could produce visual competence. Probably for most, natural selection is the only theory that meets this requirement. If so, a move toward a general selection theory has been made. The trial-and-error learning of a blind animal also can be seen as needing epistemological rationalization, and so on (Campbell, 1990).

My own strategy has been to translate the problem of knowledge into the problem of the fit (topological, map-like) between belief and referent. Thus translated, I subsume this as a subtype of the more general problem of fit between organism and environment. This fit is achieved not only by natural selection of organismic function and structure, but also by mechanisms that *vicariously* fine-tune the fit for local conditions and proximal moments of time. It is these latter that most clearly overlap with the epistemological tradition.

In my fullest presentation, I provide a tentative nested hierarchy of such knowledge processes, listing these 10 crude levels:

1. Nonmnemonic problem solving.
2. Vicarious locomotor devices (distance receptors).
3. Habit.
4. Instinct.
5. Visually supported thought.
6. Mnemonically supported thought (including computer problem solving).
7. Social vicarious exploration.

8. Language.
9. Cultural cumulation.
10. Science.

We do not have space here to examine them all, but I do want to describe a few of them. But first, an abstract statement of the dogma (Campbell, 1974b):

> A. A blind-variation-and-selective-retention process is fundamental to all inductive achievements, to all genuine increases in knowledge, to all increases in fit of system to environment.
>
> B. In such a process there are three essentials: (1) Mechanisms for introducing variation; (2) Consistent selection processes; and (3) Mechanisms for preserving and/or propagating the selected variations. Note that in general the preservation and generation mechanisms are inherently at odds, and each must be compromised.
>
> C. The many processes which shortcut a more full blind-variation-and-selective-retention process are in themselves inductive achievements, containing wisdom about the environment achieved originally by blind variation and selective retention.
>
> D. In addition, such shortcut processes contain in their own operation a blind-variation-and-selective-retention process at some level, substituting for overt locomotor exploration or the life-and-death winnowing of organic evolution. (p. 421)

Level 1. Nonmnemonic Problem Solving. This level was originally modeled after Ashby's (1952) Homeostat and conceived of a protozoan without external receptors that, when some internal receptor indicated metabolic insufficiency, would engage in a random change of locomotor direction until the metabolic deficiency disappeared. The "retention" phase of my formula was to be exemplified either by the cessation of movement, or in the continuation of movement in a straight line (rather than "retention" in memory, hence, the term *nonmnemonic*). I did allow external receptors indicating poisonous substances, which would also provoke random changes in direction of movement until the receptor activation ceased. I wrongly (Campbell, 1974b, p. 422–23) attributed this level of knowing to the paramecia. A recent review of the literature, however, finds paramecia are able to *turn in the direction* of the nourishing or away from a noxious chemical gradient, rather than having to rely on random turning (Bean, 1979) perhaps by means of a comparison of receptor intensity at the front and back of its body. (Note, however, Jensen's 1959, meticulous observations, which add much to the classic work of Jennings, 1906. Jensen describes several additional ways of changing direction, none of which seem to him to involve "comparison" or "turning toward.")

Among some still simpler unicellular organisms, prokaryotes such as bacteria and blue–green algae, we find a type of nonmnemonic problem solving that is closer to Ashby's model, in that selection of direction of movement is achieved in spite of a totally nondirectional "tumbling" (Macnab, 1979; see also Adler, 1975; Bean, 1984; Carlile, 1975). But even these organisms are surprisingly complex and subtle. The bacteria focused on by Macnab (1979), *E. Coli* and *Salmonella*, continually swim except for occasional interruptions by "tumbles" (haphazard changes in directions), in a three-dimensional random walk. In the presence of a favorable chemical gradient (increasing density for food, decreasing density for poisons), the frequency of tumbling is decreased, thus "selecting" and "retaining" a direction from among the "variations" in direction made available by "blind" tumbling. The detection of the gradient is achieved by an internal comparator generating a signal from the difference between a rapid and slow decay process of the effects of sense reception. If the faster decay product (representing the more recent sense receptor conditions) is more "favorable," tumbling is reduced in frequency. Thus, for these prokaryotes, the discriminations are based on a temporal mechanism and can be viewed as involving a primitive short-term memory. (The size of these organisms is so small as to preclude the front–back comparison of sensors, the mechanism perhaps used by the paramecia.) Are we to expect to find prokaryotes and viruses still less complex, closer to my Ashbyite model (Bean, Jensen, & Campbell, in preparation)?

One firm conclusion is that what I previously (Campbell, 1974b, 1959) offered as a single level in fact covers many discrete stages and branching alternatives. I believe it will be epistemologically valuable to spell these out. With each stage might go speculative hypotheses as to the "presumptions about" (a.k.a. "knowledge of") the external-to-the-organism world that each exhibits. In my previous analysis, I said of nonmnemonic problem solving (Campbell, 1974b): "as to presuppositions about the nature of the world (the ontology guiding epistemology), perhaps all that is assumed is spatial discontinuity somewhat greater than temporal discontinuity in the distribution of environmental substances: Moving around is judged to bring changes more rapidly than staying put. At this level the species has discovered that the environment is discontinuous, consisting of penetrable regions and impenetrable ones, and that impenetrability is to some extent a stable characteristic. The animal has 'learned' that there are some solvable problems. Already the machinery of knowing is biasedly focused upon the small segment of the world which is knowable, as natural selection makes inevitable. (p. 423). Our tumbling bacteria are still more presumptive, they "assume" that the environmental chemical gradients are "projectible" (to use Nelson Goodman's term), that they are likely to extend in the same direction in ever increasing (or ever decreasing) densities. (My Ashbyite model was much more appropriate to a world of nonsoluble solids.) We (Bean, Jensen, & Campbell, in preparation) propose to explore such a hierarchy of presumptions extending from the simplest protozoan to the primates (Jensen, 1989).

Level 2. Vicarious Locomotor Devices (distance receptors, object reification, hypostatized stable external space). To help understand this level, one may regard locomotor collisions with solid external objects as epistemologically more fundamental than the distance receptors here interpreted as *vicarious* locomotor devices, which in biological evolution have been selected and are kept tuned by natural selection. A blind man's cane and radar serve as examples. The emitted radar beam is reflected by objects having locomotor impenetrability. The wave-bands employed have been selected for this function, even though they only fallibly achieve it. X-rays that found solid objects transparent would not do. Analogously, our visual systems make us of wave-bands for which electromagnetic wave *opaqueness* coincides with locomotor *impenetrability*. (Water and fog provide paradoxical imperfection. See Campbell, 1974b, pp. 414, 423, and especially Wächtershaüser, 1987 for fuller descriptions.) Sonar and echolocation devices in animals, including the lateral-line organ in fish, provide further examples of broadcast exploration and selective reflection. (The absence of a broadcast emission process in vision complicates my fuller presentations.) All these vicarious locomotor devices are "nested within" the more fundamental, but less precise, feedback system of natural selection by whole animal survival.

(Among the levels omitted are 3: Habit, and 4: Instinct. Gary Greenberg has called my attention to the possibility of integrating these first four levels, at least, with Schneirla's 1965 approach/withdrawal theory, which will be the focus of the next Biennial Schneirla Conference.)

Level 7. Social Vicarious Exploration, and Level 8, Language. I skip to these as a final exemplar for this chapter. In the social insects, such as the ant, the trial-and-error spatial exploration of one worker provides a vicarious exploration, eliminating the need for wasteful exploratory effort on the part of other workers. An ant that has encountered food exudes a special pheromone on its way back to the nest. The other workers backtrack on that scent directly to the food supply, and if food remains plentiful, they renew the pheromone trail. If not, the trail evaporates, and the workers turn from harvesting to individualized diverse exploration.

"Vicariousness" exists here at many levels. As with the solitary ancestral species, the nourishingness of the food is vicariously and imperfectly represented by taste sensors, rather than directly through digestive metabolism. This imperfection is illustrated by both ant poison and non-nourishing saccharin. The "nestedness" is illustrated in the likelihood that natural selection is selecting mutations modifying the taste sensors in the direction of reducing the attractiveness of ant poison in ant poison-rich environments.

At the level of social sharing of food location, editing by natural selection continually tunes the "honesty" of scouts and the "trust" of followers, weeding out nests where mutations have produced scouts that sweat the "eureka" pheromone where no treasure of food has been discovered or followers who fail to trust the pheromone trail laid down by the scouts.

Concluding Comments

I had originally hoped in this chapter to integrate my levels of downward causation with the levels of vicarious selection and presumptions about the world posited in my evolutionary epistemology. I still have the intuition that so doing would clarify each, but in particular the evolutionary epistemology. Whereas I anticipate that, overall, my epistemological levels will be most akin to Dimension II of Fig. 1.1, the examples I have chosen for presentation here form a Dimension I series: Nonmnemonic problem solving for the single-celled organism, vicarious locomotor devices illustrated for multicellular organisms, and socially vicarious (linguistic) knowing for ant colonies. If limited to such a series, the two frameworks easily merge. It takes a node of selection at the multicellular individual level to keep taste sensors tuned to validly representing nourishingness. It takes a node of selection at the whole ant-nest level to keep the "honesty" and "trust" of ant scouts and followers in tune with each other. But extending this to our primary epistemological focus, ultrasocial vertebrates such as ourselves would throw my system awry and make the social contract imagery more plausible. A great deal more refinement and reconceptualization is needed.

REFERENCES

Adler, J. (1975). Chemotaxis in bacteria. In M. J. Carlile (Ed.), *Primitive sensory and communication systems: The taxes and tropisms of microorganisms and cells* (pp. 91–100). London: Academic Press.

Aronson, L. R. (1981). Evolution to telencephalic function in lower vertebrates. In P. Laming (Ed.), *Brain mechanisms of behavior in lower vertebrates* (pp. 33–58). Cambridge, England: Cambridge University Press.

Ashby, W. R. (1952). *Design for a brain*. New York: Wiley.

Baldwin, J. M. (1909). *Darwin and the humanities*. Baltimore, MD: Review. Reissued, New York: AMS, 1980.

Bean, B. (1979). Chemotaxis in unicellular eukaryotes. In W. Haupt & M. E. Feinleib, *Encyclopedia of plant physiology: New series, Vol. 7, Physiology of Movements*, (pp. 335–354). Berlin: Springer-Verlag.

Bean, B. (1984). Microbial geotaxis. In G. Colombetti & F. Lenci (Eds.), *Membranes and sensory transduction* (pp. 163–198). New York: Plenum Press.

Bean, B., Jensen, D. D., & Campbell, D. T. (in preparation). A hierarchy of levels of knowing and their associated *Umwelten*.

Boorman, S. A., & Levitt, P. R. (1980). *The genetics of altruism*. New York: Academic Press.

Bradie, M. (1986). Assessing evolutionary epistemology. *Biology and Philosophy*, *1*(4), 401–459.

Campbell, D. T. (1956a). Adaptive behavior from random response. *Behavioral Science*, *1*(2), 105–110.

Campbell, D. T. (1956b). Perception as substitute trial and error. *Psychological Review*, *63*(5), 330–342.

Campbell, D. T. (1959). Methodological suggestions from a comparative psychology of knowledge processes. *Inquiry*, *2*, 152–182.

Campbell, D. T. (1960). Blind variation and selective retention in creative thought as in other knowledge processes. *Psychological Review*, *67*(6), 380–400.

Campbell, D. T. (1965). Variation and selective retention in socio-cultural evolution. In H. R.

Barringer, G. I. Blanksten, & R. W. Mack (Eds.), *Social change in developing areas: A reinterpretation of evolutionary theory* (pp. 19–48). Cambridge, MA: Schenkman.

Campbell, D. T. (1974a). 'Downward causation' in hierarchically organized biological systems. In F. Ayala & T. Dobzhansky (Eds.), *Studies in the philosophy of biology* (pp. 179–186). Berkeley: University of California Press.

Campbell, D. T. (1974b). Evolutionary epistemology. In P. A. Schilpp (Ed.), *The philosophy of Karl R. Popper* (pp. 413–463). LaSalle, IL: Open Court Publishing. Reprinted in Campbell, 1988, pp. 393–434.

Campbell, D. T. (1983). The two distinct routes beyond kin selection to ultrasociality: Implications for the humanities and social sciences. In D. L. Bridgeman (Ed.), *The nature of prosocial development* (pp. 11–41). New York: Academic Press.

Campbell, D. T. (1987a). Neurological embodiments of belief and the gap in the fit of phenomena and noumena. In A. Shimony & D. Nails (Eds.), *Naturalistic epistemology: A symposium of two decades* (pp. 165–192). Dordrecht, Holland: D. Reidel.

Campbell, D. T. (1987b). Selection theory and the sociology of scientific validity. In W. Callebaut & R. Pinxten (Eds.), *Evolutionary epistemology: A multiparadigm program* (pp. 139–158). Dordrecht, Holland: D. Reidel.

Campbell, D. T. (1988). (E. S. Overman, Ed.), *Methodology and epistemology for social science*. Chicago, IL: University of Chicago Press.

Campbell, D. T. (1990). Epistemological roles for selection theory. In N. Rescher (Ed.), *Evolution, cognition, realism* (pp. 1–19). Lanham, MD: University Press of America.

Campbell, D. T., Heyes, C. M., & Callebaut, W. G. (1987). Evolutionary epistemology bibliography. In W. Callebaut & R. Pinxten (Eds.), *Evolutionary epistemology: A multiparadigm program* (pp. 405–431). Dordrecht, Holland: D. Reidel

Campbell, D. T., & Paller, B. T. (1989). Extending evolutionary epistemology to "justifying" scientific beliefs (A sociological rapprochement with a fallibilist perceptual foundationalism?). In K. Hahlweg & C. A. Hooker (Eds.), *Issues in Evolutionary Epistemology* (pp. 231–257). Albany, NY: State University of New York Press.

Campbell, J. H. (1986). An organizational interpretation of evolution. In D. D. Depew & B. H. Weber (Eds.), *Evolution at a crossroads: The new biology and the new philosophy of science* (pp. 133–167). Cambridge, MA: MIT Press.

Carlile, M. J. (1975). Taxes and tropisms: Diversity, biological significance and evolution. In M. J. Carlile (Ed.), *Primitive sensory and communication systems: The taxes and tropisms of micro-organisms and cells* (pp. 2–28). London: Academic Press.

Crick, F. (1966). *Of molecules and men*. Seattle: University of Washington Press.

Depew, D. D., & Weber, B. H. (1986). Innovation and tradition in evolutionary theory: An interpretive afterword. In D. D. Depew & B. H. Weber (Eds.), *Evolution at a crossroads: The new biology and the new philosophy of science*. Cambridge, MA: MIT Press.

Greenberg, G. (1988). Levels of social behavior. In G. Greenberg & E. Tobach (Eds.), *Evolution of social behavior* (pp. 137–146). Hillsdale, NJ: Lawrence Erlbaum Associates.

Jennings, H. H. (1906). *Behavior of lower organisms*. New York: Columbia University Press.

Jensen, D. D. (1959). A theory of the behavior of *Paramecium aurelia* and behavioral effects of feeding, fission, and ultraviolet microbeam irradiation. *Behavior*, *4*, 82–122.

Klee, R. L. (1984). Micro-determinism and concepts of emergence. *Philosophy of Science*, *51*, 44–63.

Kornblith, H. (Ed.). (1985). *Naturalizing epistemology* (pp. 1–17). Cambridge, MA: MIT Press.

Lewontin, R. C. (1982). Organism and environment. In H. C. Plotkin (Ed.), *Learning, development and culture: Essays in evolutionary epistemology* (pp. 151–170). New York: Wiley.

Macnab, R. M. (1979). Chemotaxis in bacteria. In W. Haupt & M. E. Feinleib (Eds.), *Physiology of movements* (pp. 310–334). New York: Springer-Verlag.

Monod, J. (1971). *Chance and necessity*. New York: Knopf.

Plotkin, H. C. (Ed.). (1982). *Learning, development and culture: Essays in evolutionary epistemology*. New York: Wiley.

Powers, W. T. (1973). *Behavior: The control of perception*. Chicago: Aldine.
Prigogine, I. (1980). *From being to becoming: Time and complexity in the physical sciences*. San Francisco: Freeman.
Pringle, W. S. (1951). On the parallel between learning and evolution. *Behaviour*, *3*, 175–215.
Schneirla, T. C. (1965). Aspects of stimulation and organization in approach–withdrawal processes underlying vertebrate behavioral development. In D. S. Lehrman, R. Hinde, & E. Shaw (Eds.), *Advances in the study of behavior* (Vol. 1, 344–412). New York: Academic Press.
Tobach, E. (1981). Evolutionary aspects of the activity of the organism and its development. In R. M. Lerner & N. A. Busch-Rossnagle (Eds.), *Individuals as producers of their development* (pp. 37–68). New York: Academic Press.
Tobach, E., & Schneirla, T. C. (1968). The biopsychology of social behavior in animals. In R. E. Cooke et al. (Eds.), *The biologic basis of pediatric practice* (pp. 68–82). New York: McGraw Hill.
Topoff, H. (1984). Levels of social integration in army-ant behavior. In G. Greenberg & E. Tobach (Eds.), *Behavioral evolution and integrative levels* (pp. 241–246). Hillsdale, NJ: Lawrence Erlbaum.
Vollmer, G. (1987). Was Evolutionare Erkenntnistheorie nicht ist. In R. Riedl & F. M. Wuketits (Eds.), *Die evolutionare erkenntnistheorie* (pp. 140–155). Berlin: Paul Parey.
Wächtershaüser, G. (1987). On light and life. In G. Radnitzky & W. W. Bartley, III (Eds.), *Evolutionary epistemology and theory of rationality, and the sociology of knowledge* (pp. 121–138). LaSalle, IL: Open Court.
Wake, D. B., & Larson, A. (1987). Multidimensional analysis of an evolving lineage. *Science*, *238*, 42–48.
Wimsatt, W. C. (1972). Teleology and the logical structure of function statements. *Studies in the History and Philosophy of Science*, *3*, 1–80.

2 Evolutionary Epistemology: Science Philosophy

Atanas Danailov and Christfried Tögel
Bulgarian Academy of Sciences

INTRODUCTION

Evolutionary epistemology has been widely discussed recently primarily because of its claim to have carried out a "Copernician turn in philosophy" (Vollmer, 1975, p. 172). The basic question evolutionary epistemology seeks to answer can be formulated as follows: What is the reason for the extensive, though not complete agreement between objective structures of the real world (i.e., physics) and the subjective structure of our knowledge of this world (i.e., mind)? Vollmer (1984) gives the following answer to this question: "Our cognitive device is a result of biological evolution. Our subjective cognitive structures match the objective structures of the world because they originated by adaptation to this world. And they are partially isomorphic to this world because otherwise we could not have survived" (p. 75). Since Kant the representation of the external world in the mind has been one of the major philosophical problems.

Before examining the philosophical position of evolutionary epistemology and its claim that a biological theory can answer a philosophical question, it would be useful to examine the prehistory of evolutionary epistemology. In doing so we will learn that neither the questions nor the answers posed by evolutionary epistemology are so surprising and new as to justify the label "Copernician revolution."

The Prehistory of Evolutionary Epistemology

The story begins with Darwin. In his N-notebook he wrote: "To study metaphysics, as they have always been studied appears to me like puzzling at astronomy without mechanics . . . the mind is a function of the body—we must bring some *stable* foundation.'' (Gruber, 1974, p. 331). There is little doubt that Darwin had an idea as to where to look for this ''*stable* foundation.'' In the M-notebook we read: ''He who understands baboons would do more towards metaphysics than Locke'' (Gruber, 1974, p. 281). This is certainly among the first claims, if not *the* first, that biology is of great importance to the solution of philosophical problems, especially epistemological ones. In the same notebook Darwin indicates the direction his thoughts are taking: ''Plato . . . says in Phaedo that our *imaginary ideas*' arise from preexistence of the soul, are not derivable from experience—read monkeys for preexistence'' (Gruber, 1974, p. 290).

This is a very interesting view of a priori synthetic judgment. During the second half of the 19th century it became so popular that the eminent historian of philosophy and founder of the neo-Kantian ''Suddeutsche Schule,'' Wilhelm Windelband (1907), could write: ''. . . this mode of explanation (intellectual progress of man being conducted by natural selection, A.D., C.T.) was frequently used by those who accept . . . the fact of a priori definitions for individual consciousness, but attempt to explain them in terms of evolution through 'acquisition' *(Erwerbung)*'' (pp. 294–295).

The group of theories Windelband had in mind was referred to as ''Biological Epistemology'' (*Biologie der Erkenntnis*,'' Kulpe, 1902; ''*Biologische Erkenntnistheorie*,'' Eisler, 1904; Jerusalem, 1919; Riehl, 1902; ''*Erkenntnistheoretisch-biologische Richtung*,'' Höffding, 1905; ''*Biologische Richtung in der Erkenntnistheorie*,'' Schwabe, 1910; ''*Evolutionsphilosphie*,'' Osterreich, 1916). Here are some quotations from those not mentioned by Campbell (1974) and Vollmer (1978) as being forerunners of evolutionary epistemology.[1]

a) In a paper called ''The theory of knowledge from an ethnopsychological point of view,'' Windelband (1875) wrote: ''If we do not believe in an uncomprehensible revelation, heredity is the easiest way to explain logical forms'' (p. 177). In a remark about this paper Steinthal (1875) said: ''. . . the hypothesis that the so-called innate ideas, the a priori categories, are products of heredity must be welcomed by ethnopsychology'' (p. 178).

b) In Ludwig Stein's (1900) book, *At the Turn of the Century* we read: ''The emergence of images (*Anschauungen*) and ideas (*Begriffe*) in the human brain . . . can be explained like all other functions and skills, inclinations and talents of the human organism: by means of selection and heredity.'' (p. 24)

c) Paul Volkmann (1910) emphasized the significance of Mach's introduction of the term *Anpassung* in the theory of knowledge and continued: ''Under the constant influences of the external necessity . . . an internal necessity of thinking evolved, which is nothing else but a copy of the external necessity . . . If one accepts this view as at least partially true, he would admit the fundamental

[1]Usually Mach, Boltzmann, Simmel, Spencer, and Baldwin are mentioned.

importance of natural science for any logic and theory of knowledge. Natural Science, indeed, would be the Archimedian point for all questions of human knowledge.'' (pp. 40–41).

d) These are only a few examples, others include: Lewes (1874–1879), Marshall (1896) and Ziehen (1913), but they demonstrate that at the turn of the century there was a well established branch of epistemology dealing with many of the same questions modern evolutionary epistemology does today. In addition, Mach, Boltzmann, Simmel, Spencer, and Baldwin are usually mentioned as well. The essential difference has to do with the extent of scientific evidence and the level of sophistication of theorizing: Evolutionary epistemology has at its disposal the synthetic theory of evolution, modern psychology, the psychophysiology of perception, psycholinguistics, and so on, while the ''Biological Epistemology'' of 100 years ago was based exclusively on the theory of Darwin.

After the First World War ''Biological Epistemology'' lost its influence, and in the 1920s and 1930s it did not play a significant role in philosophical discussions. The year 1940 marks the turning point: The biologist Konrad Lorenz accepted an appointment to the University of Konigsberg in Eastern Prussia. He was given the position once occupied by Immanuel Kant. The following year he published the paper, ''Kant's doctrine of the a priori in the light of contemporary biology,'' which is generally considered to be the first exposition of evolutionary epistemology. In this paper Lorenz (1941/1982) presented the following idea:

> Our categories and forms of perception, fixed prior to individual experience, are adapted to the external world for exactly the same reason as the hoof of the horse is already adapted to the ground of the steppe before the horse is born and the fin of the fish is adapted to the water before it hatches. (pp. 124–125)

As he further elaborated and explained,

> In our opinion, the real relationship between the thing-in-itself and the specific a priori form of its appearance has been determined by the fact that the form of appearance has developed as an adaptation to the laws of the thing-in-itself in the coping negotiation with these continuously present laws during the evolutionary history of mankind, lasting hundreds of millennia. This adaptation has provided our thought with an innate structuralization which corresponds to a considerable degree to the reality of the external world. (p. 124)

Lorenz was conscious of the fact the this is a radical new interpretation of Kant's idea of a priori knowledge. That is the reason for his warning from earlier in his paper:

> One familiar with the innate modes of reaction of subhuman organisms can readily hypothesize that the a priori is due to hereditary differentiations of the central nervous system which have become characteristic of the species, producing heredi-

> tary dispositions to think in certain forms. One must realize that this conception of the 'a priori' as an organ means the destruction of the concept: something that has evolved in evolutionary adaptation to the laws of the natural external world has evolved a posteriori in a certain sense, even if in a way entirely different from that of abstraction or deduction from previous experience. (p. 122)

We have already seen that this idea is by no means a new one. But Lorenz didn't mention any of the representatives of "Biological Epistemology" from the turn of the century. Of course, it is possible that Lorenz did not know of this body of writing. But it is difficult to believe that Lorenz was unfamiliar with the writing of such neo-Kantians as Schultz (1899, 1907), Windelband (1907), Riehl (1902), Vorlander (1903), and Simmel (1985) on the relationship between selection and knowledge, since the neo-Kantian Kurt Leider was his colleague at Konigsberg (Lorenz, 1977, p. 30). Is it really possible that Konrad Lorenz—having the university chair formerly occupied by Kant and being in the company of neo-Kantians—did not read the *Praludien* of Wilhelm Windelband, one of the basic works of the founder of the "Suddeutsche Schule"? WIndelband's paper on "Norms and Natural Laws," was published in the *Praludien*. In that paper he discussed the role of heredity and selection for the evolution of human knowledge.

This is difficult to prove as well as to believe. We propose that being both a biologist and a successor to Kant, Lorenz was intrigued by the possibility of uniting evolutionary theory and epistemology and continued the efforts of earlier thinkers in this effort.

However, Lorenz's paper about Kant's a priori remained unappreciated, and two years later he published the voluminous study on "The innate forms of possible experience" (Lorenz, 1943). This paper, however, shared the problems of the first one, and for the next 30 years Lorenz did not undertake to write further about biological epistemology. Lorenz's paper "Gestaltwahrnehmung als Quelle wissenschaftlicher Erkenntnis" (1959) cannot be considered a similar attempt, although there is a paragraph "Erkkentnistheoretische Erwagungen" ("Epistemological Considerations"), which contains ideas elaborated in Lorenz (1941) and Lorenz (1943).

Stimulated by the works of Corti (1962) and Campbell (1966), Lorenz published his famous book *Behind the mirror: A search for a natural history of human knowledge* (Lorenz, 1973). In this work he presented a detailed explanation of his ideas expressed in the early 1940s. The basic ideas are the same; even the formulations have not changed. For Lorenz there was not doubt in either period, that ". . . the categories and modes of perception of man's cognitive apparatus are the natural products of phylogeny and thus adapted to the parameters of external reality in the same way, and for the same reasons, as the horse's hooves are adapted to the prairie, or the fish's fins to the water" (Lorenz, 1977, p. 37).

The worldwide fame and notoriety that befell Lorenz after he was awarded the Nobel Prize resulted in *Behind the mirror* becoming a best seller. This book succeeded in reviving "Biological Epistemology," which, since Campbell's (1966) important paper, has been called "evolutionary epistemology."

At the end of the 1970s the "*Altenberger Kreis*" appeared. Besides Lorenz, Rupert Riedl, and others it also included a professional philosopher: Erhart Oeser. This was the first step in the institutionalization of the new biological thinking. In 1982, the 7th International Wittgenstein Symposium dedicated a whole session to evolutionary epistemology (Weingartner & Czermak, 1983). This intellectual activity led to the important international symposium on evolutionary epistemology held in Vienna in April, 1986. The opening lecture was reserved for Konrad Lorenz. His theme once again, was, "The a-priori-Problem" (Lorenz, 1987).

The Philosophical Position of Evolutionary Epistemology

Most evolutionary epistemologists would refer to their philosophical position as "hypothetical realism." This term was introduced by Donald Campbell (Campbell, 1959) and adopted by Konrad Lorenz. In his famous paper on "The perception of the 'Gestalt' as a source of scientific knowledge," Lorenz (1959) wrote that the assumption that the external world exists independently of the perceiving subject was introduced by Campbell as a *working* hypothesis. Therefore, Campbell calls the corresponding epistemological attitude *hypothetical realism*. The term *hypothesis*, however, implies the property of being able to be falsified or verified through confrontation with facts (Lorenz, 1968, p. 101).

The main theses of hypothetical realism are: (a) the hypothetical nature of our entire knowledge of reality, (b) the existence of a lawful, structured, and interconnected world, independent of mind, and (c) partial cognition of that world by means of perception, thinking, and intersubjective investigation (cf. Vollmer, 1983, p. 34). These assumptions reflect the materialistic nature of hypothetical realism. The best example of evolutionary epistemology's materialism is Lorenz's critique of idealism. He points out that doubts about the objective reality of the external world are errors with serious consequences (Lorenz, 1977, p. 28). For him idealism is an impediment to scientific investigation, because the external world is mind-independent (Lorenz, 1977, p. 26). His critique of Kant's philosophy is a materialistic one, and his arguments against idealism stem from every day experience. According to Lorenz, no one can truly believe that the reality of things in our everyday life are purely a function of our sensing processes. Every sensible person believes that the furniture in the bedroom still exists even after one has left the room (Lorenz, 1977, p. 27).

But why does our knowledge correspond to reality? What are the criteria of this correspondence? Lorenz's answers to these questions are from the perspective of scientific materialism. He is convinced that our cognitive "device" is a part of reality, and has reached its present form as a result of adapting to this reality. This is the basis of his belief that the information we have about the external world corresponds to something real (Lorenz, 1977, p. 18). He has no doubt that human knowledge results from the process of reflecting objective reality. He rejects the assumption of transcendental idealism, that there is no "reflection relation" (*Bildverhaltnis*) between thinking and reality. Thinking is the actual reproduction of reality, although a simplified reproduction (Lorenz, 1977, p. 18). In Lorenz's thinking, reflection concerns not only psychological *functions* such as thinking and cognition, but also comprises the *structure* (anatomical-morphological organization) and *behavior* of living systems, including human beings.

What is the difference between hypothetical realism and other forms of realism? For Lorenz and Vollmer "realism" is identical with "materialism." In his book *Evolutionary Epistemology* (Vollmer, 1975, 1983) Vollmer writes that each realism makes statements about the existence and the cognizability of a mind-independent external world, therefore simultaneously adopting both an epistemological and an ontological position (Vollmer, 1983, p. 34).

From this point of view, *naive* realism assumes the existence of a real world, which is structured in the way we perceive it. *Strongly critical* realism takes for granted the existence of the real world, but contends that there is a difference, in principle, between the real structures of the world and those that appear to us. Unlike these two forms of realism, *hypothetical* realism assumes the existence of the real world, which possesses certain structures, but these structures are only partially cognizable. This position is very close to *critical* realism, which accepts the existence of the real world, but claims that we are not able to perceive all its features in the way they are structured.

In defining "truth," Vollmer and other evolutionary epistemologists take the following position of "correspondence theory": *Necessary* criteria for the truth of a theory are assumed to be consistency, intersubjectivity, convergence and confirmation. These are, however, insufficient criteria (Vollmer, 1985, p. 283): Knowledge remains exclusively hypothetical because of its limits (Vollmer, 1983, p. 291).

The hypothetical realism of evolutionary epistemology has been the target of philosophical critique from various sources (Vollmer, 1985, pp. 268–323). The most serious objections come from analytical philosophy, especially from Wolfgang Stegmuller (1984). He attempts to show that neither hypothetical realism, nor the correspondence theory of truth connected with it are valid. We would like to consider only two points of Stegmuller's argument.

First, he is of the opinion that *truth* is an epistemic term, that is it has to be understood as "well-founded claimability" (*begrundete Behauptbarkeit*) or "rational acceptability" (*rationale Akzeptierbarkeit*). In the framework of hypothetical realism, a scientific theory is true if it describes reality correctly. According to Stegmuller, this has nothing to do with justification and foundation. It is this rational separation that makes hypothetical realism incomprehensible: A theory might possesses optimal epistemic characteristics such as consistency, simplicity, and so on and still not be true. In Stegmuller's view, there is no rational explanation for this situation. Hypothetical realism fails not because of contradictory, or wrong assumptions, but because of its incomprehensibility. Hypothetical realism is guided by the image of God's eye as an Archimedian point *outside* our thinking, our language or our "theorizing" (*Theoretisieren*). This point of view allows us to observe reality in itself. According to "internal realism" (Stegmuller, 1984), the order of introduction of the terms *truth* and *reality* is reversed: A theory is true if it contains all optimal epistemic characteristics, and if what it describes can be called reality.

Second, for evolutionary epistemology physics is the basic scientific discipline. The biological sciences have to continue the research program of physics. The ruling physical theory is quantum mechanics, which describes the world in the way *the observer perceives it*. Of this Stegmuller (1984) says that even quantum physics itself is not real, that is, it has no objective materialistic interpretation. In the arguments against the validity of Bell's inequality, Stegmuller sees the main reason for the failure of materialism and concludes: "For the first time in the history of science it seems to be the case that we have an empirically provable physical statement allowing us to make a judgment on the correctness of a philosophical position" (p. 26).

Answering the critique of Stegmuller and others, Vollmer defends the correspondence theory of truth by introducing the idea of "energy transfer" (*Engergieubertrag*) from cause to effect (Vollmer, 1985, pp. 104–107, 285–286). As to the second argument of Stegmuller, Vollmer (1985) writes: "Even if naive interpretation could not hold water, careful formulations are thinkable, which are compatible *both* with most of the intuitions of the realists and with quantum mechanics" (p. 286).

The Philosophical Pretensions of Evolutionary Epistemology

Can science answer philosophical questions? This is the major question to be dealt with if we want to analyze the philosophical pretensions of evolutionary epistemology, according to Vollmer. The solution to this problem depends on our understanding what philosophy is.

If we interpret philosophy as the totality of problems philosophers are concerned with, then the natural sciences can, of course, supply answers to these problems. But if we clearly demark the boundaries of the objectives and methods of philosophy on one side and those of natural science on the other, we must deny the possibility of answering philosophical questions with methods.

Curious though it may be, most philosophers and scientists believe in the successful application of science to philosophy. The shortcomings of this view are obvious (Togel, 1987): It presupposes the identification of "philosophical" and "prescientific" as well as the process of "branching-off" of scientific disciplines from philosophy (Polkarov, 1973, p. 26). But if we regard philosophy as a kind of "motherscience," from which other disciplines emanate, then this emanation must immediately be followed by an "exhaustion of philosophy or a considerable narrowing of its scope'' (Polikarov, 1973, p. 26). But this does not seem to be the case. *Philosophical* problems of scientific disciplines (philosophy of science) have always been (and will continue to be) an object of *philosophical* reflection. The solution of *scientific* problems has always been (and will continue to be) the business of the corresponding *scientific* disciplines. *It is not the case that the scientific disciplines have emanated from philosophy, but rather that the division of labor between philosophers and scientists in the last centuries has resulted in an apparent emanation.* In ancient Greece, for instance, it was entirely normal for scientists to also work in the field of philosophy. Aristotle was called the founder of psychology. But the fact that he wrote *Metaphysics* doesn't mean that his book *On sleep and waking* was also a philosophical one. Or if we consider *De anima* a philosophical work, we can likewise think of the differential calculus of Leibniz as philosophy. The history of science and philosophy has shown that the unity of philosopher and scientist in one person was the prevailing form of gaining knowledge. The fact that today this union is no longer the norm, doesn't permit us to conclude that it was not the norm at one time.

In our opinion, the existence of science is the precondition for the emergence of philosophy, because *only with the foundations of positive knowledge produced by science can we ask the philosophical (epistemological) questions about the relationship between knowledge and objective reality.* This statement draws the line between science and philosophy: The aim of science is the gaining of *knowledge about objective reality*, and the aim of philosophy (epistemology) is to investigate the *relationship between knowledge and objective reality*. The precondition of this position is the belief that there is something like ''objective reality.''

What is the significance of these considerations for our questions?

The basic question of evolutionary epistemology, ''Why do our subjective cognitive structures match with the objective structures of the world outside?,'' is not a philosophical *question*, but does, however, contain a philosophical *answer*. It expresses an explicit philosophical position concerning the relationship between knowledge and objective reality, namely, a partial isomorphism. A philosophical question in this connection would be: ''Do our subjective cognitive structures match with the objective structures of the world outside?'' And the answer of evolutionary epistemology from the point of view of hypothetical realism must be ''Yes.''

Asking the question ''Why?'' we presuppose a philosophical position. As we have seen before, the philosophical position of evolutionary epistemology is called hypothetical realism. But the ''why-question'' is not a philosophical one. The answer to this non-philosophical question is given on the basis of evolutionary theory. Hence, *evolutionary epistemology does not answer a philosophical question using results of the natural sciences, but attempts to give good scientific reasons for a certain philosophical position, that is hypothetical realism.*

Without any doubt, such an attempt is a legitimate enterprise. Evolutionary epistemology is engaged in this enterprise by trying to justify the *validity* of knowledge through the explanation of its *genesis*. This is not an entirely new approach, but the interdisciplinariness and the openness of evolutionary epistemology has refreshed the philosophical and scientific discussion on the origin and the essence of human knowledge.

ACKNOWLEDGMENTS

We thank Ethel Tobach and Donald Campbell for their fruitful comments and the useful discussions we had during the Fourth T. C. Schneirla Conference. We have tried to take their input into consideration in preparing this chapter. The authors take full responsibility for all the shortcomings still left in the chapter.

REFERENCES

Campbell, D. T. (1959). Methodological suggestions from a comparative psychology of knowledge processes. *Inquiry, 2*, 152–182.

Campbell, D. T. (1966). *Evolutionary epistemology: First variant of Campbell (1974)*. Prepared during the author's tenure as a Fellow of the Center for Advanced Study in the Behavioral Sciences.

Campbell, D. T. (1974). Evolutionary epistemology. In P. A. Schlipp (Ed.), *The Philosophy of Karl Popper* (Vol. I, pp. 413–463). LaSalle, IL: Open Court.

Corti, W. R. (1962). Das Archiv fur genetische Philosophie. Librarium. *Zeitschrift der Schweizer Bibliophilen Gesellschaft, 5*, Heft 1–2.

Eisler, R. (1904). *Worterbuch der philosophischen Begriffe*. 2. Auflage. Berlin: Mittler & Sohn.

Gruber, H. E., & Barret, P. H. (Eds.). (1974). *Darwin on man*. New York: Dutton.

Hoffding, H. (1905). *Moderne Philosophen*. Leipzig: Reisland.

Jerusalem, W. (1919). *Einleitung in die Philosophie*. (pp. 7–8). Auflage. Wien/Leipzig: Braumuller.

Kant, I. (1781). *Kritik der reinen Vernunft*. Riga: Hartknoch.

Kulpe, O. (1902). *Die Philosophie der Gegenwart in Deutschland*. Leipzig: Teubner.

Lewes, G. H. (1874–1879). *Problems of life and mind*. (5 vols.). Boston: Osgood.

Lorenz, K. (1941). Kants Lehre vom Apriorischen im Lichte gegewartiger Biologie. [Kant's doctrine of the a priori in the light of contemporary biology] *Blatter fur deutsche Philosophie, 15*, 94–125. Reprinted in H. C. Plotkin (Ed.). (1982), *Learning, development and culture: Essays in evolutionary epistemology*. Chichester, England: Wiley.

Lorenz, K. (1943). Die angeborenen Formen moglicher Erfahrung [The innate bases of learning]. *Zeitschrift fur Tierpsychologie, 5*. 235–409. Reprinted in K. H. Pribram (Ed.). (1969). *On the biology of learning*. New York: Harcourt, Brace & World.

Lorenz, K. (1959). Gestaltwahrnehmung als Quelle wissenschaftlicher Erkenntnis. *Zeitschrift fur experimentelle und angewandte Psychologie, 4*.

Lorenz, K. (1968). *Vom Weltbild des Verhaltensforschers*. Munchen: dtv.

Lorenz, K. (1977). *Behind the mirror*. New York & London: Harcourt, Brace, Jovanovich.

Lorenz, K. (1987). Das Apriori-Problem. In R. Riedl & F. M. Wuketits (Eds.), *Die evolutionare Erkenntnistheorie*. Berlin: Paul Parey.

Marshall, H. R. (1896). Consciousness and biological evolution. *Mind, 5*, 367–387, 523–538.

Osterreich, K. (Ed.). (1916). *Friedrich Uberwegs GrundriB der Geschichte der Philosophie*, (11). Auflage, Berlin: Mittler & Sohn.
Polikarov, A. (1973). *Science and philosophy*. Sofia: Publishing House of the Bulgarian Academy of Sciences.
Riehl, A. (1902). *Zur Einfuhrung in die Philosophie der Gegenwart*. Leipzig: Teubner.
Schultz, J. (1899). *Psychologie der Axiome*.
Schultz, J. (1970). *Die drei Welten der Erkenntnistheorie*.
Schwabe, O. (1910). *Die biologische Richtung in der neueren Erkenntnistheorie*. Hamburg: Schuler & Jeve.
Simmel, G. (1895). Uber eine Beziehung der Selectionslehre zur Erkenntnistheorie. *Archiv fur systematische Philosophie, 1*, 34–45.
Stegmuller, W. (1984). Evolutionaire Erkenntnistheorie, Realismus und Wissenschaftstheorie. In R. Spaeman, P. Koslowski, & R. Low (Eds.), *Evolutionstheorie und menschilches Selbstverstandnis*. Weinheim: Verlag Chemie.
Stein, L. (1900). *An der Wende des Jahrhunderts*. Freiburg i. Br.
Steinhal, H. (1875). Zusatz zu Windelband. *Zeitschrift fur Volkerpsychologie und Sprachwissenschaft, 8*, 178.
Vollmer, G. (1983). *Evolutionare Erkenntnistheorie*. 3. Auflage. Stuttgart: Hirzel.
Vollmer, G. (1984). Mesocosm and objective knowledge. In F. M. Wiketits (Eds.), *Concepts and Approaches to Evolutionary Epistemology*. Dordrecht, Holland: D. Reidel.
Vollmer, G. (1985). *Was konnen wir wissen? Bd. 1: Die Natur der Erkenntnis*. Stuttgart: Hirzel.
Vorlander, K. (1903). *Geschichte der Philosophie*. 2. Band. Leipzig: Durr'sche Buchhandlung.
Weingartner, P., & Czermak, H. (Eds.). (1983). Epistemology and philosophy of science. *Proceeding of the 7th International Wittgenstein Symposium*, Kirchberg a. Wechsel, Austria, 1982. Wien: Holder-Pichler-Tempsky.
Windelband, W. (1875). Die Erkenntnislehre unter dem volkerpsychologischen Gesichtspunkte. *Zeitschrift fur Völkerpsychologie und Sprachwissenschaft, 8*, 166–178.
Windelband, W. (1907). *Praludien*. Tubingen: Mohr.
Ziehen, T. (1913). *Erkenntnistheorie auf psychopysiologischer physikalischer Grundlage*. Jena: Gustav Fischer.

3 A Developmental Contextual Critique of Evolutionary Epistemology

Richard M. Lerner
The Pennsylvania State University

As the "normal" social or behavioral scientists "toil in the vineyards" of their discipline, it is rare for them to unearth a concept that, at one and the same time, spans several sciences, possesses substantial philosophical significance, and has profound social policy implications. Simply, most behavioral scientists are used to dealing with concepts and issues rather insulated from implications for other domains of scholarship or for the world of policy, application, and politics. Yet, in the subject area encompassed by the concept of "evolutionary epistemology," the scientist encounters a topic having considerable importance for philosophy, for science, and for policy (cf. Chorover, 1979).

In this chapter I emphasize what I see as some of the core, and yet most controversial and problematic, scientific features of evolutionary epistemology. I also touch on aspects of evolutionary epistemology's philosophical and social policy implications. It is useful to presage here the key points I make in regard to the links among science, philosophy, and policy.

In my view, the concept of evolutionary epistemology is based on, at best, an incomplete depiction of evolutionary processes. The problematic conception of evolutionary process found in evolutionary epistemology involves an over-reliance on natural selection—on adaptation—as *the* mechanism in phylogenetic change; there is also an exclusion of emphasis on the organism's activity across its ontogeny and on the dynamic interactions between an active organism and a changing context. The incomplete conception of evolutionary process found in evolutionary epistemology leads to a philosophically paradoxical compartmentalization of organisms and the "natural" world (cf. Clark, 1984; Prigogine, 1971, 1978). In turn, and despite evolutionary epistemologists' claims about eschewing the "incomprehensible arrogance" (Lorenz, 1941) of seeing the cognitive system of humans as qualitatively discontinuous from the knowledge systems of other species, the view of evolutionary processes found in evolutionary epistemology does lead to the belief that there *are* privileged instances of cognitive systems. These privileged instances are those particular structural, functional, or content dimensions of a cognitive system that, as a consequence of natural selection, are fit, are adaptive. It is these dimensions that have been selected by nature, and they take precedence over any other, non-naturally selected process.

How may we understand what may seem, at first blush, to be a curious and contradictory use of a notion of non-natural selection within a conception that advocates the exclusivity of natural selection in shaping organismic characteristics? It is important to consider that evolutionary epistemological thinking is often coupled with a predetermined epigenetic (Gottlieb, 1976) view of the basis of the presence of "adaptive" cognitive components in the ontogeny of an organism (e.g., Lorenz, 1965). Hence there is a stress on instinctually, genetically determined shaping of these cognitive components, and a denial of the role of organism–environment interactions across ontogeny as being involved in the genesis of these components.

When there is a link between evolutionary epistemology and a predetermined epigenetic view of development, it is but a short step from natural selection of desirable cognitive components (viz, the adaptive ones) to a societally superimposed selection process when, for whatever reasons, nature has erred and allowed genetic material carrying the nonadaptive (viz, degenerate) cognitive components to be insinuated into adaptive human stock (e.g., see Lorenz, 1940). I illustrate that such a line of reasoning can lead to social policies that may range from sterilization to genocide (Chorover, 1979; Kalikow, 1978; Lifton, 1986).

Any scientific theory may be misused, be it theory based or not based on firm philosophical and empirical grounds. However, I argue that when the thinking about adaption found in evolutionary epistemology is coupled with a predetermined epigenetic view of ontogeny, such misuse may be more probable. Therefore, I suggest that when this coupling occurs in the work of a person having particular political ideologies in mind, such misuse moves away from the realm of probability to the domain of certainty.

Clearly, it is a major understatement to say that I take exception to key features of evolutionary epistemology. Moreover, I am aware I am being provocative in raising the last, social policy points. As a "normal" scientist I have had training to be rational, unemotional, and polite in professional discourse. However, it is because evolutionary epistemology is such a profoundly and unusually important conception and does in my view so dramatically include potentially dire social policy implications, that I have cast these introductory remarks in the preceding manner. As a consequence, therefore, it is important that I explain in some detail the reasoning involved in my criticisms. This reasoning derives most centrally from a probabilistic epigenetic (Gottlieb, 1976) view of ontogenetic change. Within the literature of developmental psychology, this conception has been termed a developmental contextual view of life-span development (Lerner, 1986; Lerner & Kauffman, 1985). To present the critique of evolutionary epistemology afforded by developmental contextualism, it is useful to first present the features of evolutionary epistemology that engender my objections.

FEATURES OF EVOLUTIONARY EPISTEMOLOGY

In a field having as rich a literature as is the case in regard to evolutionary epistemology, it is not surprising to realize that there exist differences in definitions and in the interpretation of key concepts (e.g., compare Campbell, 1977, with Richards, 1977). Keeping this point in mind it is nevertheless possible to offer what may be a generally subscribed to definition of the field itself. For instance, according to Tennant (1983): "Evolutionary epistemology is concerned not so much with an account of the 'evolutionary' progression of scientific theories as with the origin and development of cognitive structures and abilities, in a long phylogenetic line stretching from amoeba to man" (p. 33).

Tennant's definition underscores, first, the phyletic continuity in cognitive structure, content, and function on which Lorenz (1941) insists, and for which there is a long history in the writings of evolutionists that dates back to Darwin (1871). As noted by Bitterman (1960), Darwin, stressing both functional and structural continuity, believed that phylogenetic differences in intelligence where differences "of degree and not of kind," and that even in respect to cognitive abilities that might seem qualitatively discontinuous and unique to humans (e.g., language) Darwin claimed that this "half-art and half-instinct" ability was clearly marked by "the stamp of its gradual evolution" (Darwin, 1871, pp. 105–106).

Tennant's view that the origin and development of the cognitive system derives from a continuous evolutionary process is a position not only having a long scientific history but is, as well, a stance that is shared by other contemporary scholars of evolutionary epistemology. For example, Danailow and Togel (1987) note Vollmer's (1984) view that the main idea in evolutionary epistemology is that: "Our cognitive device is a result of biological evolution. Our subjective cognitive structures match with the objective structures of the world because they originated by adaptation to this world. And they are partially isomorphic to this world because we could not have survived" (p. 75).

This definition, which stresses the continuity of Tennant (1983) and/or the gradualism of Darwin (1871), emphasizes two other features of evolutionary epistemology. First, in stressing that the cognitive system is a product of the "objective" world, Vollmer depicts a unidirectional, materially realistic conception of causation (Clark, 1984). The cognitive system is the consequent of antecedent events in the physical world, a world that is independent of these contingent thought processes. No possibility exists for the cognizing organism to act on the physical world, for if it influences the very context that is shaping it, it therefore engages in bidirectional, or reciprocal, organism-physical context causation. There is no role admitted for ontogenetically changing organism-context interactive patterns to provide a means for influence on the physical world and, as such, for eventual evolutionary change. Instead, we are left, according to Vollmer (1984), with a view not unlike that forwarded by Mach (1905/1976) near the beginning of this century; that is, that in an individual's life, thoughts adapt *to* the "facts" of the objective world because of the biological fitness of knowledge and cognitive functions (Danailow & Togel, 1987).

This correspondence between Mach and Vollmer leads me to note the second feature of evolutionary epistemology emphasized by Vollmer (1984), and I should add by other contemporary advocates of evolutionary epistemology (e.g, Campbell, 1977). The view of evolutionary progression found in evolutionary epistemology is one of adaptation—the shaping of a structure and/or function by natural selection, with a resulting fitness between that structure and/or function and the demands for survival. Indeed, it is the process of natural selection that is seen as the unidirectional, materially realistic means by which the objective world causes the cognitive system. Thus, we hear evolutionary epistemologists discuss the mechanisms of natural selection—involving, for instance, "variation and selective retention" (Campbell, 1977, p. 504), or the "*objective pull of nature*" (Annis, 1982, p. 205)—in order to emphasize that a key "external constraint on an epistemic practice is its survival value" (Annis, 1982, p. 205), its adaptive character.

In essence, then, evolutionary epistemology represents a phylogenetic continuity view of the origin and development of the cognitive system and, as such, is a view in which ontogenetic activity by the organism is not given causative status either in the ontogenetic shaping of the cognitive system or in altering the context within which the organism develops. Instead, this context unidirectionally shapes, by natural selection, the organism's cognitive system, with the outcome being the presence of adaptive, that is, fit, cognitive systems.

It is in this way that the cognitive system, or in fact any organismic system, has its origin. It is this view of its origin that determines the character of the ontogenetic development of the cognitive system. The organism's cognitive system, being a complete consequent of antecedent natural selection processes, cannot be influenced in any causal, and/or qualitatively distinct, way by the ontogenetic activity of the organism, for instance, its reciprocal, or dynamic, interactions with the physical context. Instead, an organism's ontogenetic change is completely provided for by its evolutionary past and is, therefore, conveyed in totality by physically materialistic entities—genes—that are construed to be evolutionarily shaped blueprints for an organism's development. Thus, genes, and the processes they are believed to invariantly produce (that is, maturational ones), are seen to predetermine the course of the ontogenetic development of the cognitive system.

It is because of the implicit presence of such reasoning that evolutionary epistemology is coupled, virtually necessarily in my view, with what Gottlieb (1975, 1983) has labeled *predetermined epigenesis*. Within this perspective, hereditarily fixed and determined intrinsic processes (i.e., maturational ones) are seen to essentially unfold; although environmental or experiential variables may speed up or slow down these progressions, they can do nothing to alter the sequence or quality (e.g., the structure) of these hereditarily predetermined changes (e.g., see Gesell, 1946; Hamburger, 1957). In other words, this view (Gottlieb, 1983):

> as it applies to behavior, means that the development of behavior in larvae, embryos, fetuses, and neonates can be explained entirely in terms of neuromotor and neurosensory maturation (i.e., in terms of proliferation, migration, differentiation, and growth of neurons and their axonal and dendritic processes). In this view, factors such as the use or exercise of muscles, sensory stimulation, mechanical agitation, environmental heat, gravity, and so on, play only a passive role in the development of the nervous system. Thus, according to predetermined epigenesis . . . structural maturation determines function, and not vice versa. (p. 11).

Victor Hamburger's (1957) organismic position epitomizes this view: "the architecture of the nervous system and the concomitant behavior patterns result from self-generating growth and maturations processes that are determined entirely by inherited, intrinsic factors, to the exclusion of functional adjustment, exercise, or anything else akin to learning" (p. 56).

In summary, these features of evolutionary epistemology, although certainly not the only components of the perspective one might mention, are the key ones that make the perspective quite problematic. If continuity, adaptation, unidirectional causation, and therefore organismic developmental irrelevance are not exclusive or even plausible, features of the origin and development of cognitive systems, then evolutionary epistemology as a scientifically viable and coherent perspective is, at best, severely limited. There are excellent reasons to believe that such limitation exists, and it is precisely because the concepts of continuity, adaptation, unidirectionality, and organismic developmental irrelevance are not supportable. It is useful to turn to a discussion of what may be a major evolutionary process involving other than adaptation.

THE ADAPTATIONIST PROGRAM AND THE CONCEPT OF EXAPTATION

The concept of adaptation is at the core of evolutionary epistemological thinking. Yet the meaning of this concept is not as clear as one would like for a term carrying such a heavy philosophical and scientific load. Gould and Vrba (1982) note that there are two current meanings in the biological literature for the term *adaptation*: First, "a feature is an adaptation only if it was built by natural selection for the function it now performs," whereas the second definition conceives of "adaptation in a static or immediate way as any feature that enhances current fitness, regardless of its historical origin" (p. 5).

To illustrate the use of these two distinct conceptions of the term, Gould and Vrba cite Williams (1966) as illustrating the first definition. Williams contended that one should speak of adaptation only when one can "attribute the origin and perfection of this design to a long period of selection for effectiveness in this particular role" (p. 6). Williams distinguishes further between adaptations and their functions (i.e., their operations), on the one hand, from fortuitous, "chance" effects, on the other. By *effect* Williams (1966) means simply something caused or produced (e.g., by a function), a result, or a consequence (e.g., of a function).

The writings of Bock (1967, 1979, 1980), in contrast to those of Williams, illustrate the second definition of adaptation. For instance, Bock (1979) indicates that "an adaptation is . . . a feature of the organism . . . which interacts operationally with some factor of its environment so that the individual survives and reproduces" (p. 39).

Gould and Vrba (1982) believe that a confusion therefore exists about a concept central in evolutionary theory, a conflict arising as a consequence of "subsuming different criteria of historical genesis and current utility under a single term" (p. 5). Indeed, Darwin (1859) himself may have seen this potential confusion: "The sutures in the skulls of young mammals have been advanced as a beautiful adaptation for aiding parturition, and no doubt they facilitate, or may be indispensable for this act; but as sutures occur in the skulls of young birds and reptiles, which have only to escape from a broken egg, we may infer that this structure has arisen from the laws of growth, and has been taken advantage of in the parturition of the higher animals" (p. 197).

Whereas, Darwin saw the necessity of unfused sutures in the skulls of young mammals, he eschewed—as Williams (1966) would argue later, one indeed should eschew—labeling the unfused sutures as adaptations. Darwin did this because the unfused sutures "were not built by selection to function as they now do in mammals" (Gould & Vrba, 1982). But if the unfused sutures are not adaptations, what are they? Clearly, the taxonomy of form in relation to fitness lacks a term, and this is the omission Gould and Vrba seem to rectify.

Following Williams (1966), they (Gould & Vrba, 1982) define:

> as an *adaptation* any feature that promotes fitness and was built by selection for its current role (criterion of *historical genesis*). The operation of an adaptation is its *function*. . . . We may also follow Williams in labelling the operation of a useful character not built by selection for its current role as an *effect*. (We designate as an effect only the usage of such a character, not the character itself . . .). . . . But what is the unselected, but useful character itself to be called? Indeed it has no recognized name. . . . Its space on the logical chart is currently blank.

> We suggest that such characters, evolved for other usages (or for no function at all), and later "coopted" for their current role, be called *exaptations*. . . . They are fit for their current role, hence *aptus*, but they were not designed for it, and are therefore not *ad aptus*, or pushed towards fitness. They owe their fitness to features present for other reasons, and are therefore *fit (aptus)* by *reason of (ex)* their form, or *ex aptus*. Mammalian sutures are an exaptation for parturition. Adaptations have functions, exaptations have effects. The general, static phenomenon of being fit should be called aptation, not adaptation. (The set of aptations existing at any one time consists of two partially overlapping subsets: the subset of adaptations and the subset of exaptations. This also applies to the more inclusive set of aptations existing through time. (p. 6)

A clear implication of Gould and Vrba's revised terminology is that not all instances of fitness are adaptations; that is, not all features of an organism's structure and function that are aptational have this character as a consequence of being shaped by natural selection for this character. Such a possibility, if supported, would serve to weaken what Gould and Lewontin (1979) have labeled the *adaptationist program*, the position that a feature's current aptational character implies historical shaping by natural selection for that character.

Lewontin (1981) has discussed the adaptationist "program" and its conventional use of the concept of adaptation. As do Gould and Vrba, Lewontin sees problems with the "adaptationist program" view of adaptation; in essence, he sees the view as deficient because it ignores the active, constructive role the organism plays in its own adaptation: The organism shapes the context to which it adapts, and hence there is a reciprocal, multilevel (i.e., dynamic interactional) relation between organism and context.

For example, summarizing the literature pertaining to the character of the environment to which organisms adapt, Lewontin and Levins (1978) stress that reciprocal processes between organism and environment are involved in human evolution, supporting the view that human functioning is one source of its own evolutionary development. Lewontin and Levins (1978) state that:

> "the activity of the organism sets the stage for its own evolution. . . . the labor process by which the human ancestors modified natural objects to make them suitable for human use was itself the unique feature of the way of life that directed selection on the hand, larynx, and brain in a positive feedback that transformed the species, its environment, and its mode of interaction with nature." (p. 78).

Clearly, then, Lewontin is formulating a position of evolution and of its link to development that is quite distinct from the predetermined epigenetic perspective found with evolutionary epistemology. Instead, Lewontin is forwarding a conception that, as I argue next, is consistent with a probabilistic epigenetic, or developmental contextual, perspective. His commitment to such a perspective is made clear in his specific criticisms of biologists committed to the adaptationist program. Lewontin (1981) notes that such biologists:

define adaptation as some kind of partial match with the external world. This is the view of a static, or at least independent, outer world that poses fixed problems to which the organism (or population) responds by fitting itself to the preexisting external condition. . . . But life is not like that. Organisms . . . by their own life activities determine which aspects of the outer world make up their environment. Organisms change the environment by their activities. They transduce physical signals from the environment into new physical forms within themselves. Species do not, in general, solve preset environmental "problems" by gathering information and responding appropriately. Rather they "construct" environments. The problem is that the concept of adaptation has been extended metaphorically from its valid domain of describing individual, short-term, goal-directed behavior to other levels. Individual organisms can be observed to be adaptive machines as they steer around obstacles, chase prey, defend themselves from attack, or push their roots around stones. But it is pure metaphor, ideologically molded by the progressivism and optimalism of the nineteenth century, to describe numbers of chromosomes, patterns of fertility, migrations, and religious institutions as adaptations. It is this kind of error that led to the now discredited descriptions of lemming migrations and mass suicides as adaptive responses to crowding. It is not simply that some evolutionary process can be described as nonadaptive, but that the entire framework is in question. Whether we look at the fossil record or at living species, we do not see them as adapting, but as adapted. But how can that be? How is it that, if evolution is a process of *adapting*, organisms always seem to be *adapted*? It may be more illuminating to see organisms as *changing*, and, in the process, as reconstructing the elements of the outer world into a new environment that is sufficient for their survival. To accept the metaphor of adaptation for human culture is to restrict the possible explanations of culture to a relatively narrow range. The price of metaphor is eternal vigilance. (p. 245)

As a consequence of their similar objections to the adaptationist program, Gould and Vrba (1982) contend that if nonapatation "is about to assume an important role in a revised evolutionary theory, then our terminology of form must recognize its cardinal evolutionary significance—cooptability for fitness" (p. 7). In other words, recognition of the potential presence of exaptative, aptative features leads one to recognize that previously nonaptative (note, *not* preadaptive) features may be present and may be coopted for fitness—a recognition that provides a key for plasticity in evolutionary processes.

To illustrate this basis for plasticity, it is useful to consider exaptation pertinent to the features of microevolution. This illustration of exaptation indicates how this concept may account for a feature of the genome that, to those committed to an adaptationist program, might appear anomalous. Gould and Vrba (1982) point out that:

for a few years after Watson and Crick elucidated the structure of DNA, many evolutionists hoped that the architecture of genetic material might fit all their presuppositions about evolutionary processes. The linear order of nucleotides might be the beads on a string of classical genetics: one gene, one enzyme; one nucleotide substitution, one minute alteration for natural selection to scrutinize. We are now, not even 20 years later, faced with genes in pieces, complex hierarchies of regulation and, above all, vast amounts of repetitive DNA. High repetitive, or satellite, DNA can exist in millions of copies; middle-repetitive DNA, with its tens to hundreds of copies, forms about one quarter of the genome in both *Drosophilo* and *Homo*. What is all the repetitive DNA for (if anything)? How did it get there? (p. 10)

Some of the repeated DNA may be conventional adaptations, selected for a role in regulation (for example, the repeated copies may bring previously separated parts of the genome into new, aptative interrelation). However, there is too much repetitive DNA for such direct adaptation to account for all of it. A second, traditional (i.e., adaptationist program-oriented) basis for the presence of so much repeated DNA has been forwarded. This suggestion is that repetitive DNA exists because it is needed for *future* evolution; that is, it exists to provide for a "flexible future"; for instance, nonused, redundant copies are free to alter because their adaptative product is still being produced by the remaining DNA copies (e.g., Cohen, 1976; Kleckner, 1977). However, this second argument is teleological because it permits future needs to determine present circumstances.

Whereas Gould and Vrba (1982) believe that future uses are quite significant consequences of repeated DNA, the potential future use cannot be held to empirically determine the prior status of the genome. In turn, the concept of exaptation capitalizes on the idea that repeated DNA may indeed have a significant future use but does so without recourse to teleological, "final cause" explanations. And in making these contributions, the concept of exaptation furthers understanding of how features of the genome provide a basis for plastic microevolutionary processes. Gould and Vrba (1982) explain that:

defenders of the second tradition understand how important repetitive DNA is to evolution, but only know the conventional language of adaptation for expressing this conviction. But since utility is a future condition (when the redundant copy assumes a different function or undergoes secondary adaptation for a new role), an impasse in expression develops. To break this impasse, we might suggest that repeated copies are nonapted features, available for cooptation later, but not serving any direct function at the moment. When coopted, they will be exaptations in their new role (with secondary adaptive modifications if altered).

What then is the source of these exaptations? According to the first tradition, they arise as true adaptations and later assume their different function. The second tradition, we have argued, must be abandoned. A third possibility has recently been proposed (or rather, better codified after previous hints): perhaps repeated copies can originate for no adaptive reason that concerns the traditional Darwinian level of phenotypic advantage (Orgel & Crick, 1980; Doolittle & Sapienza, 1980). Some DNA elements are transposable; if these can duplicate and move, what is to stop their accumulation as long as they remain invisible to the phenotype (if they

> become so numerous that they begin to exert energetic constraint upon the phenotype, then natural selection will eliminate them)? Such "selfish DNA" may be playing its own Darwinian game at a genic level, but it represents a true nonaptation at the level of the phenotype. Thus, repeated DNA may often arise as a nonaptation. Such a statement in no ways argues against its vital importance for evolutionary futures. When used to great advantage in that future, these repeated copies are exaptations. (p. 11)

In other words, and crucial for a synthesis of micro and macroevolutionary processes, Gould and Vrba (1982) believe that there exists an "enormous pool" of nonaptations and that this pool must be the source, the "reservoir," of most evolutionary flexibility. They note that:

> We need to recognize the central role of "cooptability for fitness" as the primary evolutionary significance of ubiquitous nonaptation in organisms. In this sense, and at its level of the phenotype, this nonaptive pool is an analog of mutation—a source of raw material for further selection.
>
> Both adaptations and nonaptations, while they may have non-random approximate causes, can be regarded as randomly produced with respect to any potential cooptation by further regimes of selection. Simply put: all exaptations *originate* randomly with respect to their effects. Together, these two classes of characters, adaptations and nonaptations, provide an enormous pool of variability, at a level higher than mutations, for cooptation as exaptations [and provide for] . . . the flexibility of phenotypic characters as a primary enhancer of or damper upon future evolutionary change. Flexibility lies in the pool of features available for cooptation (either as adaptations to something else that has ceased to be important in new selective regimes, as adaptations whose original function continues but which may be coopted for an additional role, or as nonaptations always potentially available). The paths of evolution—both the constraints and the opportunities—must be largely set by the size and nature of this pool of potential exaptations. Exaptive possibilities define the internal contribution that organisms make to their own evolutional future. (pp. 12, 13)

To summarize, the concept of exaptation, and the limitations it imposes for the notion of adaptation as the sole process by which evolution occurs, present formidable conceptual and empirical problems for the evolutionary epistemology perspective per se and for the implications of this perspective for philosophy, science, and policy. The existence of exaptive processes indicates that evolution is considerably more plastic than evolutionary epistemology, in general, or predetermined epigenesis, in particular, would imply. This plasticity is highlighted by those objections to the adaptationist program that involve the specification of the causal role played by the developing organism, and especially by its dynamic interactions with its context, in influencing the course of evolution. It is this developmental contextual interaction that is at the core of the key theoretical alternative to predetermined epigenesis; that is, probabilistic epigenesis. It is useful to turn now to a discussion of this alternative view.

PROBABILISTIC EPIGENESIS, DEVELOPMENTAL CONTEXTUALISM, AND THE ROLE OF THE CONCEPT OF BEHAVIORAL NEOPHENOTYPES IN EVOLUTIONARY CHANGE

As described by Lewontin (1981), it is possible to view the organism as other than just the host of its evolutionarily provided genes, and it is likewise possible to view the importance of the organism's activity across ontogeny as more than just the maturationally predetermined unfolding of hereditarily fixed progressions. The key alternative view is one that sees biological and contextual factors as reciprocally interactive; as such, developmental changes are probabilistic in respect to normative outcomes due to variation in the timing of the biological, psychological, and social factors that provide interactive bases of ontogenetic progressions (e.g., Schneirla, 1957; Tobach, 1981). This view has been labeled as *probabilistic epigenetic* by Gottlieb (1983), and developed by him (Gottlieb, 1976) and earlier by Schneirla (1956, 1957) and Tobach and Schneirla (1968). The term *probabilistic epigenesis* was used by Gottlieb (1970): "to designate the view that the behavioral development of individuals within a species does not follow an invariant or inevitable course, and, more specifically, that the sequence or outcome of individual behavioral development is probable (with respect to norms) rather than certain" (p. 123).

Moreover, Gottlieb (1970) explains that this probable, and not certain, character of individual development arises because:

> Probabilistic epigenesis necessitates a bidirectional structure-function hypothesis. The conventional version of the structure-function hypothesis is unidirectional in the sense that structure is supposed to determine function in an essentially nonreciprocal relationship . . . The bidirectional version of the structure-function relationship is a logical consequence of the view that the course and outcome of behavioral epigenesis is probabilistic: it entails the assumption of reciprocal effects in the relationship between structure and function whereby function (exposure to stimulation and/or movement of musculoskeletal activity) can significantly modify the development of the peripheral and central structures that are involved in these events. (p. 123)

Thus, within the probabilistic epigenetic view of development, or—as it is termed within the area of life-span developmental psychology—the developmental contextual (Lerner, 1986; Lerner & Kauffman, 1985) view of development, variables from biological, psychological, and social contextual levels of analysis bidirectionally interact to provide the bases of behavior and development. It is important to stress that levels in this literature are conceived of as integrative organizations; that is (Novikoff, 1945):

> the concept of integrative levels recognizes as equally essential for the purpose of scientific analysis both the isolation of parts of a whole and their integration into the

> structure of the whole. It neither reduces phenomena of a higher level to those of a lower one, as in mechanism, nor describes the higher level in vague non-material terms which are but substitutes fro understanding, as in vitalism. Unlike other "holistic" theories, it never leaves the firm ground of material reality . . . The concept points to the need to study the organizational interrelationships of parts and whole. (p. 209)

Moreover, Tobach and Greenberg (1984) stress that: "the interdependence among levels is of great significance. The dialectic nature of the relationship among levels is one in which lower levels are subsumed in higher levels so that any particular level is an integration of preceding levels . . . In the process of integration, or fusion, *new* levels with their own characteristics result" (p. 2). If the course of human development is the product of the processes involved in the "fusions" or "dynamic interactions" (Lerner, 1978, 1979, 1984) among integrative levels, then the processes of development are more plastic than often previously believed (cf. Brim & Kagan, 1980).

Lewontin (1981) has indicated what such plasticity in development may mean for altering the course of evolution. More recently, Gottlieb (1987) has provided a more detailed, and quite intriguing, discussion of the role of plastic developmental functioning in shaping evolutionary change.

Although biologists such as Garstang (1922), de Beer (1930), and Goldschmidt (1933) have argued previously that developmental changes may lead to evolution, they believed as well that a genetic change or a mutation was necessary to create the developmental changes. Gottlieb (1987), however, argues for an evolutionary pathway in which ontogenetic development leads to evolutionary change and, quite significantly, where "genetic change is a secondary or tertiary consequence of enduring behavioral changes brought about by nongenetic alterations of species-typical development" (p. 267). Gottlieb's conception draws on a notion introduced by Kuo (1967), of behavioral neophenotype, that is, a behavioral innovation, or ontogenetic novelty, made possible by the plasticity of the organism and of its probabilistic, dynamic interactions with its context.

Gottlieb contends that a behavioral neophenotype is the likely first step in an evolutionary sequence that proceeds from behavioral change, to morphological change, to genetic change. More specifically, the emergence of a behavioral neophenotype encourages new environmental relationships that, in turn, bring out latent possibilities for morphological and physiological changes. Gottlieb (1987) notes that somatic mutation, cytoplasmic alteration, or change in gene regulation may at this point take place as well; however, an alteration of structural genes need not take place in this secondary stage of the process. However, a change in genes or in gene frequency does occur in the third stage, wherein as a consequence of long-term geographic or behavioral isolation (i.e., separate breeding populations) such alteration takes place.

Because of the plasticity that exists in organisms (and especially ones with larger relative brain size, such as humans; Gottlieb, 1987), a plasticity textured by the probabilistic, dynamic interactions they have across ontogeny with their context, an evolutionary pathway is created that is inconsistent with the conception of evolutionary change found in evolutionary epistemology and in the associated predetermined epigenetic view of organism change. Thus, on the basis of both this contribution to evolution by organism development, and by the implications of the concept of exaptation, we may conclude that the key features of evolutionary epistemological thinking are severely scientifically limited. Evolution processes are not therefore just comprised of phylogenetically continuous changes that, by virtue of the antecedent and independent effects of the physical world, shape via natural selection particular cognitive structures, functions, or contents to be adaptive. As a consequence, the particular set of cognitive features present in a person, a social group, or a culture cannot be judged as contributing to or diminishing the survival of the human species by virtue of the adaptationist assertion that the features have or have not been shaped and selected for fitness. This conclusion leads me to some final comments, ones that return me to a discussion of the possibly dire social implications of evolutionary epistemology.

CONCLUSIONS: POTENTIAL IMPLICATIONS OF EVOLUTIONARY EPISTEMOLOGY FOR SOCIAL POLICY

All too frequently in the history of scientific discourse about the role of biology in human characteristics and development—be these features cognitive, personological, or even moral ones—the argument noted above has been made, that is, that particular features of human functioning have been naturally selected for fitness. Indeed, as I indicated at the outset of this chapter, when the kind of adaptionist thinking found in evolutionary epistemology is coupled, as it seems necessarily to be, with predetermined epigenesis, there emerges an implicit—and sometimes explicit—belief in the existence of privileged cognitive systems: Those cognitive systems, which by virtue of natural selection, have been shaped to be adaptive are therefore the privileged, desirable ones. And, although it is philosophically curious and logically contradictory to assert that whereas, on the one hand, *all* cognitive systems are the outcomes of antecedent and independent natural phenomena acting through the process of natural selection, and that there are, on the other hand, outcomes for which nature has erred, it has often been the case that such an assertion is made. Indeed, the philosophical and logical problem is then compounded when, on the basis of such an assertion, there is an appeal to introduce a social policy or program that will "correct" nature's mistake—a suggestion seemingly impossible to implement *if*, as the evolutionary epistemologist would have it, human systems such as cognitive ones are only the consequences of a materially realistic world and as such do not have feedback loops too, or dynamic interactions with, this world.

Nevertheless, such arguments have been made; and social programs have been implemented either because of, or at least using the justification of, such views. Here, then, lies the potentially dire societal consequences of the problematic positions found in evolutionary epistemology. If some organisms possess the privileged systems and others, for whatever reasons (e.g., mutation, genetic anomaly) do not, then because of the coupling of adaptationist and predetermined epigenetic thinking, the only recourse to a society aspiring to be fit, to survive, is to ensure that the carriers of the nonfit, or degenerate, genes—which of course transmit the inadequate structures, functions, or abilities—not be allowed to reproduce. What is seen to be at stake is nothing less than the survival of the privileged, of the fit, of the species.

Lest you think that I am carrying these implications too far and am constructing an imaginary scenario that could never be produced in any modern culture, let me illustrate the argument I am describing by citing the work of a Nobel laureate who is one of the leading advocates both of evolutionary epistemology and of predetermined epigenesis (Kalikow, 1978, 1987; Lifton, 1986). Writing in National Socialist Germany in 1940, Konrad Lorenz (translated by Eisenberg, 1972) asserted:

> The only resistance which mankind of healthy stock can offer . . . against being penetrated by symptoms of degeneracy is based on the existence of certain innate schemata. . . . Our species-specific sensitivity to the beauty and ugliness of members of our species is intimately connected with the symptoms of degeneration, caused by domestication, which threaten our race. . . . Usually, a man of high value is disgusted with special intensity by slight symptoms of degeneracy in men of the other race. . . . In certain instances, however, we find not only a lack of this selectivity . . . but even a reversal to being attracted by symptoms of degeneracy. . . . Decadent art provides many examples of such a change of signs. . . . The immensely high reproduction rate in the moral imbecile has long been established . . . . This phenomenon leads everywhere . . . to the fact that socially inferior human material is enabled . . . to penetrate and finally to annihilate the healthy nation. The selection for toughness, heroism, social utility . . . must be accomplished by some human institution if mankind, in default of selective factors, is not to be ruined by domestication-induced degeneracy. *The racial idea as the basis of our state has already accomplished much in this respect*. The most effective race-preserving measure is . . . the greatest support of the natural defenses. . . . We must—and should—rely on the healthy feelings of our Best and charge them with the selection which will determine the prosperity or the decay of our people. (italics added)

Although switching his scientific focus from Jews to precocial birds after the Nazis lost the Second World War, Lorenz maintained an interest in human instincts and continued to view them in a manner consistent with a presupposition that they could not be moderated by the kinds of interactions described by the notion of probabilistic epigenesis. In his book, *On Aggression*, Lorenz (1966) argued that aggression was instinctual in humans, that social conflict was an innate component of the human behavioral repertoire, and that there was an instinctual "militant enthusiasm" among youth. Although in this book Lorenz suggests that civilization must find methods for ritualizing and containing aggressive instincts, he is basically arguing that humans must make the best of an inevitably negative situation (Schneirla, 1966); that is, he sees humans as universally and unalterably destructively aggressive.

The social-policy implications of this instance of Lorenz's views about "instinct" were perhaps best presented in a review of *On Aggression* written by Schneirla (1966): "It is as heavy a responsibility to inform man about aggressive tendencies assumed to be present on an inborn basis as it is to inform him about "original sin," which Lorenz admits in effect. A corollary risk is advising societies to base their programs of social training on attempts to inhibit hypothetical innate aggressions, instead of continuing positive measures for constructive behavior.'" (p. 16).

As I asserted at the opening of this chapter, the "normal" scientist may not be accustomed to dealing with a concept having profound multidisciplinary and societal ramifications. Yet, we have been cautioned by Schneirla (1966), as we have been alerted by Lewontin (1981), that we must remain aware of such far-reaching implications. In fact, because, as I noted earlier, any scientific theory may be misused, the need for such vigilance seems continual. As a consequence, we should standardly include, as part of our evaluation of the significance and meaning of a theory or scientific body of work, a consideration of its societal implications. Whereas no scientific work should ever be censored or repressed as a consequence of its implications, the coupling of societal and scientific significance evaluations is, in my view, legitimate. Indeed, the evaluation of societal implications should be approached with the same theoretical rigor and methodological precision that we associate with any component of science. Indeed, if we do not take these steps we may be creating the conditions selecting against our own very existence.

ACKNOWLEDGMENTS

The preparation of this chapter was supported in part by a grant to Richard M. Lerner form the William T. Grant Foundation and by NIMH Grant MH39957. I thank Donald H. Ford, Kathryn Hood, Jacqueline V. Lerner, Robert Plomin, Alexander von Eye, and Fred W. Vondracek for critical comments on previous drafts of this manuscript.

REFERENCES

Annis, D. B. (1982). Epistemology naturalized. *Metaphilosphy, 13*, 201–208.

Bitterrman, M. E. (1960). Toward a comparative psychology of learning. *American Psychologist, 15*, 704–712.

Bock, W. J. (1967). The use of adaptive characters in avian classification. *Proceeding of the XIV International Ornithology Congress.*

Bock, W. J. (1979). A synthetic explanation of macroevolutionary change—a reductionistic approach. *Bulletin of the Carnegie Museum of Natural History, 13*, 20–69.

Bock, W. J. (1980). The definition and recognition of biological adaptation. *American Zoologist, 20*, 217–227.

Brim, O. G., Jr., & Kagan, J. (1980). Constancy and change: A view of the issues. In O. G. Brim, Jr. & J. Kagan (Eds.), *Constancy and change in human development*. Cambridge, MA: Harvard University Press.

Campbell, D. T. (1977). Comment on "the natural selection model of conceptual evolution." *Philosophy of Science, 44*, 502–507.

Chorover, S. L. (1979). *From genesis to genocide*. Cambridge, MA: The MIT Press.

Clark, A. J. (1984). Evolutionary epistemology and ontological realism. *The Philosophical Quarterly, 34*, 482–490.

Cohen, S. N. (1976). Transposable genetic elements and plasmid evolution. *Nature, 263*, 731–738.

Danailow, A., & Togel, C. (1987). *On the history of evolutionary epistemology*. Unpublished manuscript, Institute of Philosophy, Bulgarian Academy of Sciences, Bulgaria.

Darwin, C. (1859). *Origin of species*. London: John Murray.

Darwin, C. (1871). *The descent of man and selection in relation to sex*. (Vol. 1). London: Murray.

de Beer, G. (1930). *Embryology and evolution*. Oxford: Clarendon Press.

Doolittle, W. F., & Sapienza, C. (1980). Selfish genes, the phenotype paradigm, and genome evolution. *Nature, 284*, 601–603.

Eisenberg, L. (1972). On the human nature of human nature. *Science, 176*, 123–128.

Garstang, W. (1922). The theory of recapitulation: A critical restagement of the biogenetic law. *Journal of the Linnean Society of London, Zoology, 35*, 81–101.

Gesell, A. L. (1946). The ontogenesis of infant behavior. In L. Carmichael (Ed.), *Manual of child psychology*. New York: Wiley.

Goldschmidt, R. (1933). Some aspects of evolution. *Science, 78*, 539–547.

Gottlieb, G. (1970). Conceptions of prenatal behavior. In R. Aronson, E. Tobach, D. S. Lehrman, & J. S. Rosenblatt (Eds.), *Development and evolution of behavior: Essays in memory of T. C. Schneirla*. San Francisco: Freeman.

Gottlieb, G. (1976). The roles of experience in the development of behavior and the nervous system. In G. Gottlieb (Ed.), *Neural and behavioral specificity: Studies on the development of behavior and the nervous system* (Vol. 3). New York: Academic Press.

Gottlieb, G. (1983). The psychobiological approach to developmental issues. In M. M. Haith & J. J. Campos (Eds.), *Handbook of child psychology: Infancy and biological bases* (4th ed.). New York: Wiley.

Gottlieb, G. (1987)., The developmental basis of evolutionary change. *Journal of Comparative Psychology, 101*, 262–271.

Gould, S. J., & Lewontin, R. C. (1979). The spandrels of San Marco and the Panglossian paradigm: A critique of the adaptationist programme. In J. Maynard Smith & R. Holliday (Eds.), *The evolution of adaptation by natural selection*. London: Royal Society of London.

Gould, S., & Vrba, E. (1982). Exaptation: A missing term in the science of form. *Paleobiology, 8*, 4–15.

Hamburger, V. (1957). The concept of development in biology. In D. B. Harrid (Ed.), *The concept of development*. Minneapolis: University of Minnesota Press.

Kalikow, T. J. (1978). Konrad Lorenz's "Brown past": A reply to Alec Nisbett. *Journal of the History of the Behavioral Sciences, 14*, 173–179.

Kalikow, T. J. (1987). Konrad Lorenz as evolutionary epistemologist: The problem of intentionality. In A. Shimony & D. Nails (Eds.), *Naturalistic epistemology* (pp. 119–144). Dordrecht: D. Reidel.

Kleckner, N. (1977). Translocatable elements in prokaryotes. *Cell, 11*, 11–23.

Kuo, Z. Y. (1967). *The dynamics of behavior development*. New York: Random House.

Lerner, R. M. (1978). Nature, nurture, and dynamic interactionism. *Human Development, 21*, 1–20.

Lerner, R. M. (1979). A dynamic interactional concept of individual and social relationship development. In R. Burgess & T. Huston (Eds.), *Social exchange in developing relationships*. New York: Academic Press.

Lerner, R. M. (1984). *On the nature of human plasticity*. New York: Cambridge University Press.

Lerner, R. M. (1986). *Concepts and theories of human development* (2nd ed.). New York: Random House.

Lerner, R. M., & Kauffman, M. B. (1985). The concept of development in contextualism. *Developmental Review, 5*, 309–333.

Lewontin, R. C. (1981). On constraints and adaptation. *Behavioral and Brain Sciences, 4*, 244–245.

Lewontin, R. C., & Levins, R. (1978). Evolution. *Encyclopedia, Vol. 5, Divino-Fame*. Torino, Italy: Einaudi.

Lifton, R. J. (1986). *The Nazi doctors*. New York: Basic Books.

Lorenz, K. (1940). Durch Domestikation verursachte Storungen arteigenen Verhaltens. *Zeitschrift fur angewandte Psychologie und Charakterkunde, 59*, 2–81.

Lorenz, K. (1941). Kant's Lehre vom Apriorischen im Lichte gegenwartiger Biologie. *Blatter fur deutsche philosophie, 15*, 94–125.

Lorenz, K. (1965). *Evolution and modification of behavior*. Chicago, University of Chicago Press.

Lorenz, K. (1966). *On aggression*. New York: Harcourt, Brace, & World.

Mach, E. (1976). *Enkenntnis und Irrtum [Knowledge and error]*. Boston: D. Reidel Publishing Company. (Original work published 1905)

Novikoff, A. B. (1945). The concept of integrative levels of biology. *Science, 62*, 209–215.

Orgel, L. E., & Crick, F. H. C. (1980). Selfish DNA: The ultimate parasite. *Nature, 284*, 604–607.

Prigogine, I. (1971). Interpretations of life and mind. In A. Greene (Ed.), *Essays around the problem of reduction*. New York: Humanities Press.

Prigogine, I. (1978). Time, structure, and fluctuation. *Science, 201*, 777–785.

Richards, R. J. (1977). The natural selection model of conceptual evolution. *Philosophy of Science, 44*, 494–501.

Schneirla, T. C. (1956). Interrelationships of the innate and the acquired in instinctive behavior. In P. P. Grasse (Ed.), *L'instinct dans le comportement des animaux et de l'homme*. Paris: Mason et Cie.

Schneirla, T. C. (1957). The concept of development in comparative psychology. In D. B. Harris (Ed.), *The concept of development*. Minneapolis: University of Minnesota Press.

Schneirla, T. C. (1966). Instinct and aggression. *Natural History, 75*, 16ff.

Tennant, N. (1983). In defense of evolutionary epistemology. *Theoria, 49*, 32–48.

Tobach, E. (1981). Evolutionary aspects of the activity of the organism and its development. In R. M. Lerner & N. A. Busch-Rossnagel (Eds.), *Individuals as producers of their development: A life-span perspective*. New York: Academic Press.

Tobach, E., & Greenberg, G. (1984). The significance of T. C. Schneirla's contribution to the concept of levels of integration. In G. Greenberg & E. Tobach (Eds.), *Behavioral evolution and integrative levels*. Hillsdale, NJ: Lawrence Erlbaum Associates.

Tobach, E., & Schneirla, T. C. (1968). The biopsychology of social behavior of animals. In R. E. Cooke & S. Levin (Eds.), *Biologic basis of pediatric practice*. New York: McGraw-Hill.

Vollmer, G. (1984). Mesocosm and objective knowledge: On problems solved by evolutionary epistemology. In F. M. Wuketits (Ed.), *Concepts and approaches in evolutionary epistemology* (pp. 69–121). Dordrecht & Boston: Reidel.

Williams, G. C. (1966). *Adaptation and natural selection*. Princeton: Princeton University Press.

4 Development of Experience and Action: Levels of Integration in Human Functioning

Seymour Wapner and Jack Demick
Clark University

INTRODUCTION

Rather than clarifying the concept of or projecting general questions concerning evolutionary epistemology, the theme of this volume, we have chosen to characterize briefly our own perspective to the study of mind and to illustrate concretely how this approach defines central issues and shapes research questions for the empirical investigation of experience and action in human functioning.

The theoretical perspective presented here—a variation of Werner's (1957a, 1957b) organismic-developmental approach—bears some striking similarities to Piaget's (1952, 1960, 1971) genetic epistemology. These similarities include, for example, the central place of such notions as genesis, the active organism, structurization of experience, and comparative methodology. However, there are differences. Most predominant is the restriction of Piagetian theory largely to human cognition, to scientific-logical analysis, and to its focus on the "epistemological subject." In contrast, our perspective attempts to encompass affective and valuative as well as cognitive processes and to focus on the "vital subject" in the context of the everyday life situation.[1] Further, our approach is linked to Schneirla's (1949, 1957, 1972) insofar as it is comparative, stressing concepts of qualitatively different levels of integration in human and, to a lesser extent, in infrahuman functioning. Thus, in keeping with Franklin's (1981) analysis, our

[1]Piaget's (1960) relatively brief treatment of the vital subject is mainly confined to viewing affectivity as a factor in accelerating or retarding formation of cognitive structures; he did, however, stimulate work on the relation between cognition and affectivity, such as that of Anthony (1957), Decarie (1963), and Wolff (1960) (cf. Wapner, 1978).

approach represents not only a psychology of mind, but also psychologies of organism, of person, and of situation.

From our viewpoint (e.g., Wapner, 1981, 1986, 1987; Wapner & Demick, 1987a; Wapner, Kaplan, & Cohen, 1973; Wapner, Cohen & Kaplan, 1976; Wapner, Ciottone, Hornstein, McNeil & Pacheco, 1983), the concept of levels of organization/integration (e.g., Feibleman, 1954; Herrick, 1949; Novikoff, 1945a, 1945b; Schneirla, 1949, 1957, 1972; von Bertalanffy, 1950) is central to the understanding of the operations (here, the experience and action) of the vital, living human being in all his/her complexity in the everyday life situation. By employing the levels concept in human functioning, our work complements those approaches that emphasize the levels concept in other animals (e.g., Aronson, Tobach, Lehrman, & Rosenblatt, 1972; Greenberg & Tobach, 1984).

The thrust of this chapter is to exemplify our modes of approaching—both theoretically and empirically—the levels concept in human functioning. We have made more progress in shaping some ways in which this general problem can be conceptualized and subjected to empirical inquiry than in uncovering the key relationships we ultimately seek. Nevertheless, we proceed on the assumption that a brief presentation of these small beginnings takes us one step closer to our ambitious goal.

THE PERSPECTIVE

The perspective we describe here has a long history and distinguished lineage dating back to 1926, when Werner first published his comparative psychology of mental development.

A synoptic description of our current conceptualization of this approach is provided in Fig. 4.1 (see Wapner, 1987).[2]

Organism-in-Environment System

The rectangle in Fig. 4.1 depicts the *organism-in-environment system*[3] *as the unit to be analyzed* (cf. Dewey & Bentley, 1949). Organism and environment are

---

[2]The formulations systematically build on: (1) Heinz Werner's seminal work, *Comparative Psychology of Mental Development* (originally published in German in 1926), which was later recast as the organismic-developmental approach (Werner, 1957a, 1957b; Werner & Kaplan, 1963); (2) Werner & Wapner's (1949, 1952) sensory-tonic field theory of perception (See Wapner, Cirillo, & Baker, 1969, 1971, and Wapner & Werner, 1957, 1965); (3) Werner and Kaplan's now classic work on *Symbol formation* (1963); (4) the writings of colleagues and students of Werner who applied his approach to a variety of subject matters in psychology (Wapner & Kaplan, 1983); (5) Kaplan's efforts to integrate Wernerian conceptualization with that of Kenneth Burke in "genetic-dramatism" (Kaplan, 1983); and finally, (6) Wapner's recent work in environmental psychology with colleagues and students (e.g., Kaplan, Wapner, & Cohen, 1976; Wapner, 1977, 1981, 1987; Wapner, Ciottone, Hornstein, & McNeil, 1983; Wapner, Kaplan, & Cohen, 1973).

[3]The organism-in-environment system represents the most general case insofar as it is comprised of human and nonhuman organisms and their environments.

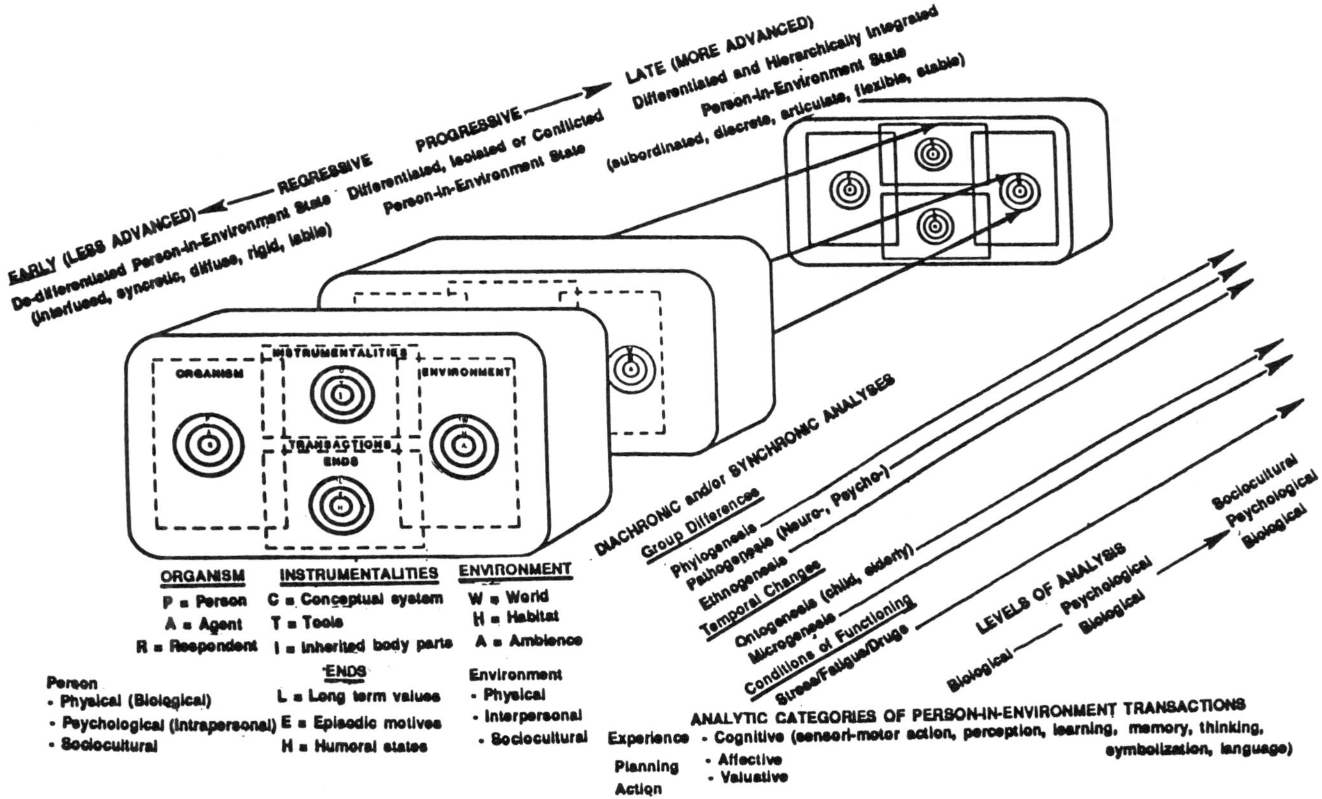

FIG. 4.1 A holistic, developmental, systems approach to person-in-environment functioning (adapted from Wapner, 1986).

linked by *transactions*, which include experience (cognitive, affective, valuative) and action. The organism and its environment are viewed as *structural* components. The *dynamic* aspects of the system, or the transactions of organism and environment, are characterized in terms of ends (goals) and instrumentalities (means). The organism-in-environment system is analyzed in formal terms. Focusing on the characteristic *structure* of the system involves analysis of more or less differentiated parts or subsystems that are more or less integrated with one another in specifiable ways. Focusing on the *dynamics* of the system entails the determination of the *means* by which a characteristic structure is achieved or maintained.

Levels of Integration

The organism-in-environment is further differentiated in terms of *levels of integration* (Wapner, Kaplan, & Cohen, 1973). At the least complex level, we refer to the organism-in-environment as *respondent-in-ambience*. A respondent transacts with the ambience in terms of such behavior as kineses, taxes, and orientation (Fraenkel & Gunn, 1961), which are fixed, reflex-like responses of the organism as a whole to ambient stimulation (e.g., the optomotor turning response in the orientation of the beetle to the movement of light). For such systems, means and ends are difficult to distinguish. They include biological drives and sensory and locomotor systems.

At the next level, the organism-in-environment is referred to as an *agent-in-habitat* (e.g., a chimpanzee solving the problem of reaching for food located behind an obstruction). The agent transacts with the habitat (e.g., social and nonsocial objects and resources) in terms of seeking satisfaction of episodic motives and incentives (e.g., food, shelter, mate, avoidance of danger). The agent-in-habitat achieves these ends through the use of such instrumentalities as tools (e.g., mechanisms for enhancing the senses, extending functions of body parts).

At the most complex level, we characterize the organism-in-environment as a *person-in-world*. A *person* (living entity *qua* acculturated human being) transacts with the environment called world, which consists of multiple sociocultural scenes (e.g., via such sociocultural objects as a university; political, economic, military, religious, kinship rules and sanctions) both individually and collectively. For such systems, the unique goals are comprised of long-term values and norms; the unique instrumentalities include use of conceptual systems (e.g., symbols, language, myth, scientific concepts) and planning.

Organism-in-environment processes or functions may similarly be categorized in terms of *levels of integration*; that is, we assume that the respondent-in-ambience system functions only at the biological level, whereas the agent-in-habitat system functions on the psychological level as well. For the person-in-world system, functioning at the biological (e.g., breathing, perspiring), the psychological (e.g., sensorimotor action, perceiving, thinking), and the sociocultural levels (i.e., living by governmental rules and regulations, having a recorded history) are similarly related in a contingency fashion; that is, functioning at the sociocultural level requires functioning at the psychological and biological levels and psychological functioning requires biological functioning. The levels differ qualitatively and functioning at one level is not reducible to functioning on the prior, less complex level because we assume that higher level functioning does not substitute for but rather integrates and transforms lower level functioning.

Although our assumptions apply to organisms-in-environments generally, our research for the most part has been restricted to human beings; hence, we exemplify our remaining assumptions in terms of the person in his/her world or environment, hereafter referred to as the *person-in-environment*. In this system, the person is comprised of physical/biological, intrapersonal, and sociocultural aspects and the environment is comprised, analogously, of physical, interpersonal, and sociocultural aspects.

Holistic Assumption

The person-in-environment system operates as a unified whole so that a disturbance in one part affects other parts and the totality. This *holistic* assumption also holds for functioning within a given level (See Fig. 4.1). For example, on the psychological level, such part-processes as the cognitive aspects of experience and action (including sensorimotor functioning, perceiving, thinking, learning, imagining, symbolizing) as well as the affective and valuative aspects of experience and action operate contemporaneously and in an integrated fashion in the normal functioning adult (cf. Wapner & Demick, 1987b). The person, his/her environment, and the transactions (experience and action) between them involving means and ends are all interdependent.

Teleological Directedness

The inclusion of dynamic components—means and ends—points to the proposition that the transactions of the person-in-environment system are not chaotic or random; rather the system is directed from some initial state of person-in-environment functioning toward some optimal end state. Here, the approach operates as a framework for describing processes (Werner & Kaplan, 1956) and thereby is comfortable with teleonomy (cf. Ayala, 1970; Campbell, 1974; Hofstadter, 1941; Thompson, 1987).

Structuring

That the person actively structures and transacts with the environment in terms of goals achieved by a variety of means conforms to the constructivist or interpretationist position of the approach (cf. Lavine, 1950). According to this position, human beings construct objects of perception and thought and thereby actively contribute to the cognitive process.

Equilibration Tendencies

Organism-in-environment systems are characterized as operating in a dynamic equilibrium. Ongoing organism-in-environment relations may be disturbed or perturbed by a change in the organism, in the environment, or in both. This may make for dramatic qualitative changes in relations among system components (organism, environment, transactions, means, ends). Moreover, following perturbation of the organism-in-environment system, reestablishment of a new dynamic equilibrium or ongoing state directed toward accomplishing goals is assumed to take place.

This process is readily demonstrated in our research dealing with critical person-in-environment transitions across the life-span. For example, to take an extreme case, we have reviewed reports of the course of changes following the impact of a sudden shock to the person-in-environment system (e.g., through tornado, earthquake, atomic bomb) and found that powerful disruptions in system components and the relations between them are followed by a new dynamic equilibrium with new goals being structured and pursued in constructive ways, as defined by the developmental principle described next (Wapner, 1983).

Developmental Analysis

Our view of development transcends the boundaries within which the concept of development ordinarily is applied. For most psychologists, development is restricted to ontogenesis. Here, in contrast, we assume that "*development* becomes a manner of looking at phenomena, a new way of representing phenomena, rather than merely a particular phenomenon in itself" (Kaplan, 1966, p. 622). We are concerned with studying both processes of formation (integration) and of dissolution (disintegration) as living organisms attempt to seek an optimal relationship with their environments.

Components (organism; environment), relations among components (e.g., means–ends), and part-processes of organism-in-environment systems (e.g., cognition in the person-in-environment system) are assumed to be developmentally orderable in terms of the *orthogenetic principle* (Kaplan, 1966, 1967; Werner, 1957a; Werner & Kaplan, 1956). The orthogenetic principle defines development in terms of the degree of organization attained by a system. The more differentiated and hierarchically integrated a system is in terms of its parts, and of its means and ends, the more highly developed it is said to be. For example, in the person-in-environment system, optimal development entails differentiated and hierarchically integrated experience and action, with flexibility, freedom, self-mastery, and the capacity to shift from one mode of person-in-environment relationship to another as required by goals, by the demands of the situation, and by the instrumentalities available (e.g., Kaplan, 1966; Wapner, 1977, 1981, 1987; Wapner, Ciottone, Hornstein, McNeil, & Pacheco, 1983; Wapner, Kaplan, & Ciottone, 1981).

The orthogenetic principle as applied to the person-in-environment system is characterized in Fig. 4.1. From left to right, there is a progression from "dedifferentiated person-in-environment state" to a "differentiated, isolated or conflicted person-in-environment state" to the most advanced, "differentiated and hierarchically integrated person-in-environment state."

Further, under *temporal changes*, we include *ontogenesis* (changes from conception to maturity and old age) and *microgenesis* (formation of a percept, a thought, a judgment, or a concept). Under *group differences*, we include: *phylogenesis* (developmental comparison and contrast of activities or adaptation manifest by typical members of different species), *pathogensis* (pathology—organic and functional—studied on the basis of a corollary of progression, viz., regression, disintegration, dissolution, breakdown, or failure to develop) and *ethnogenesis* (developmental comparison of changes in mental functioning that occur during the history of humankind).

Finally, *conditions of functioning* can also be viewed developmentally. For example, sleeping/waking, alcohol/normal, LSD-25/normal, stress/optimal represent extremes varying respectively from less to more advanced with respect to the orthogenetic principle.

SOME RELATIONS WITHIN AND AMONG LEVELS: EMPIRICAL RESEARCH AND FUTURE DIRECTIONS

Although marked strides have been made in delineating the significance of the levels of integration concept—especially with respect to animal functioning—there is still a need to specify some relations between and among levels as well as to illustrate empirical methodologies designed to assess such relations. From the perspective delineated here, we attempt to specify some of these relations both within a given level (here, psychological) as well as between and among levels. Further, we specify some of these relations at given stages of development (e.g., in the normal functioning adult) as well as through comparison of developmentally ordered series (e.g., ontogenesis, pathogenesis, temporary states of functioning). We draw our examples both from work conducted some time ago and from more recent work. Our formulations for empirical study derive from two facets of our perspective (viz., its function as ". . . both a framework for describing processes and a framework for formulating empirically testable hypotheses") (Werner & Kaplan, 1956, p. 869).

Relations Within Psychological Level

In keeping with the holistic aspect of our approach, we attempt to provide direct demonstrations of the interrelations among part-processes and among cognitive processes at the psychological level of the person-in-environment system.

Cognition and Affect. Some interrelationships between *cognition and emotion* are illustrated in a series of studies that demonstrated the impact of *danger on experience of space and time* (Langer, Wapner, & Werner, 1961; Langer, Werner, & Wapner, 1965; Werner & Wapner, 1955a). In these studies (Werner & Wapner, 1955a), subjects looked at the precipitous edge of a theater stage or a stairwell from a distance. They were then requested to walk up to or were moved toward the precipice while blindfolded. The most striking findings were that, depending on the task or goal, perceived distance, perceived time, and speed of walking diminished as subjects approached a dangerous locale. Thus, there is equivalence of the experience of distance, duration, and speed under conditions of danger, and which aspect of experience is expressed or channeled (principle of vicarious channelization) depends on the task or goal to which the person is directed (Werner & Wapner, 1949, 1952).

Many other studies conducted in the Clark laboratories have also dealt with interrelationships between cognition and affect. Taken together, these experiments have suggested that affect may impact cognition, not only under conditions of danger, but also under conditions of success/failure (Wapner, Werner & Krus, 1957), psychopathological states of mania and depression (Rosenblatt, 1956), emotional states induced by tranquilizers and antidepressants (Krus, Resnick, & Raskin, 1966), and positive and negative affective relationships to others (Isaac, 1958).

The analysis of the *development* of relations between cognition and affect can be readily approached through use of the orthogenetic principle. For example, Werner (1957a) described use of physiognomic perception by less developmentally advanced persons as perception fused with affect (e.g., the child's exaggeration of the size of her favorite doll; the schizophrenic's looking fearfully at a swinging door saying 'the door is devouring me').

More specifically, with respect to the changing relations of these processes during *ontogenesis*, we would take a different approach from those who argue that cognitive functioning is (e.g., Lazarus, 1981, 1982) or is not (e.g., Zajonc, 1980) necessary for affective functioning. Rather, we might, for example, ask: In adulthood, is the optimal relationship between cognition and affect characterized by hierarchic integration between the two? Further, is there flexibility whereby one process is subordinated to the other dependent on the demands of the situation (e.g., emotion may be subordinated to cognition in psychotherapeutic endeavors, whereas cognition may be subordinated to emotion in artistic endeavors)? In older adulthood, do the relations between cognition and affect regress to a more de-differentiated state similar to that of childhood?

With respect to *pathogenesis*, our perspective leads us to ask whether there are parallels among such emotional states as schizophrenia (psychopathology), dementia (neuropathology), and the early stages of ontogenesis. Further, is the pattern of relations between cognition and affect formally similar in the schizophrenic, the dementing individual, and the young child? For example, rather than having a differentiated and integrated relation, is the pattern de-differentiated (e.g., characterized by physiognomic perception where cognition and affect are fused), differentiated and isolated (e.g., where affect is split off from cognition), or differentiated and in conflict (e.g., where intellectually one feels one way and emotionally another)?

Cognition and Valuation. Studies dealing with the interrelationships between cognition and valuation date back to the 'New Look' in perception of the 1950s, which demonstrated the role of subjective factors such as attitudes, values, and needs in the perception of objects and their attributes (cf. Allport, 1955; Blake & Ramsey, 1951; Bruner & Goodman, 1947; Zener, 1949a, 1949b). From the perspective described here, the part-processes of cognition and valuation in the aesthetic appreciation of painting were found to be integrated in different ways dependent on duration of exposure to the stimulus (painting); that is, at brief exposures, only perceptual processes (e.g., directional properties of lateral organization) operated in the evaluation (e.g., preference, balance, dynamics) of the painting; at longer exposures, higher level conceptual processes (e.g., symbolic interpretation of content) additionally affected evaluation (Freimuth & Wapner, 1979). A second study, concerning *developmental* relationships between cognition and valuation, demonstrated that a candy bar was generally drawn larger than a neutral object (piece of wood), and that this size-value effect decreased during ontogenesis (Baker, Rierdan, & Wapner, 1974).

Affect and Valuation. Our tripartite focus on cognition, affect, and valuation suggests that studies concerning the interrelationships between emotion and valuation as well as among cognition, affect, and valuation are worthy of systematic investigation. It is possible that such studies have been largely neglected because of the general notion that affect and valuation are positively correlated. However, there may be cases in which highly valued objects, states, or events are negatively cathected, or vice versa. Our perspective suggests that the examination of this relationship and its links to cognition represent areas worthy of future systematic investigation both generally and developmentally.

Relations Among Cognitive Processes: General Considerations. Relations among cognitive processes within the psychological level may also be analyzed in terms of levels of integration. As outlined by Werner (1957a) and elaborated by Wapner (1969), *sensorimotor* (direct, external motoric manipulation of objects), *perceptual* (directiveness towards properties of things or people in external space), and *conceptual-symbolic* (the manipulation of symbols representative of objects) processes represent the three strata of cognitive functioning. Sensorimotor operations are assumed to emerge first, followed by perceptual operations and, finally, by conceptual-symbolic operations. An operation that emerges at later stages of development does not supplant an operation already present but is assumed to be hierarchically integrated with it; that is, the mature individual is assumed to have at his/her disposal the early- as well as the late-appearing functions.

Relations Among Cognitive Processes: Some Types. Some leads for studying relationships among processes derive from studies that have already been conducted exploring cognitive operations in terms of such relations as *vicarious, supportive, antagonistic, correspondent*. A *vicarious* relationship implies that the utilization of one operation substitutes for the use of another. A *supportive* relationship implies that simultaneously occurring cognitive operations facilitate one another to make for greater understanding and efficiency in achievement of ends. An *antagonistic* relation is one where the operation of one function lessens the operation of another. A *correspondent* relation occurs when one level of functioning parallels another.

Studies on these types of relationships with adults were conducted fairly early in our experimental program. Each is exemplified here. *Vicarious* relationships between simultaneously operating cognitive functions were demonstrated in studies that showed that, under conditions of inhibition of motor expression, there was an increase in perceived autokinetic motion and vice versa (Goldman, 1953). Further, the introduction of motor involvement significantly decreases perceptual movement as measured by verbal responses to pictorial material (Krus, Werner, & Wapner, 1953). Thus, there is evidence in keeping with vicarious channelization (viz., an inverse relation between sensorimotor and perceptual activity).

Supportive relations have been examined in experimental studies using lapse of meaning and perceptual recognition of words. For example, Miller (1963) found that "lapse-of-meaning," which occurs with verbal repetition of a word ("push"), is delayed significantly when there is sensorimotor activity, simultaneously with its repetition consonant with its meaning (pushing activity). The sensorimotor activity operates supportively insofar as it serves to prolong the meaning. An analogous relationship holds for threshold of perceptual recognition: Sensorimotor activity consonant with the meaning of the words lowers the recognition threshold (Dowling, 1965; Dowling, Werner, & Wapner, 1960).

Developmental changes in lapse of meaning utilizing verbal repetition of the word, both in the context of concrete activity and its absence, were studied in children and adults. There is evidence that: (1) younger age groups, generally, show less stability in maintaining meaning with repetition of a word; (2) if concrete activity is congruent with meaning, it enhances stability of meaning (i.e., retards onset of lapse of meaning); and (3) concrete activity is differentially more efficacious in the youngest and the oldest groups, implying that there is not a unitary process at all age levels (Wapner & Rand, 1967).

In addition to these changes ordered developmentally with respect to ontogenesis, there is a parallel phenomenon with respect to groups ordered in terms of neuropathology; that is, supportive relations are also evident in cases where a brain-injured person with agnosia can begin to understand meaning of words when the letters of a word are traced sensorimotorically (Weisenburg & McBride, 1935). Another parallel phenomenon has also been reported in child development, where sensorimotor activity has been shown to improve form discrimination (Zaporozhets, 1965).

Evidence for an *antagonistic* relation between cognitive operations was found in a learning situation utilizing developmentally ordered groups in terms of ontogenesis as well as psychopathology (Switzer, 1961). The particular experimental procedure involved requesting subjects to use externalized sensorimotor operations, e.g., pointing while learning a sequence of stimuli, (a series of colored lights flashed in boxes located at different positions). It was found that, in contrast to such operations when they occurred spontaneously, externalized sensorimotor operations under these "forced" conditions functioned to disrupt performance. Thus, for superimposed sensorimotor operations to act supportively in learning situations, the subject must be developmentally advanced enough to be able to integrate the different operations.

Correspondence between perceptual and conceptual operations was also explored empirically (Teft, Wapner, Werner, & McFarland, 1964). The general hypothesis was tested that if an extent (perceptual operation) is demarcated by two numbers (from 1 to 9), the perception of extent varies depending on the numerical interval (conceptual operation) used. It was found that a visual extent demarcated by a smaller numerical interval appeared equal to a visual extent demarcated in a darkroom by a greater numerical interval when the former extent was in fact physically smaller relative to the latter. This correspondence between perceptual and conceptual cognitive levels is assumed to occur because both a conceptual numerical interval and visual (physically determined) length have the organismic characteristic of expansiveness.

Relations Among Cognitive Processes: Sensorimotor Action and Perception. A second series of studies (Wapner, Cirillo, & Baker, 1969) describing relations among different cognitive levels of integration is based on the principle that the complexity of organization of a system as a whole increases with the emergence of higher levels (cf. Aronson, 1987; Feibleman, 1954; Tobach, 1987). The cognitive processes, from less to more complex, chosen were: (1) *sensorimotor action* in maintaining balance; (2) *perceptual objectification* in localization of an object (e.g., perception of verticality, perception of body position); and (3) *experienced relations between percepts* (e.g., localization of one's own body and of objects out-there).

The first level concerns *maintaining the body erect and balanced*. Here, burdens on the body, pulls in one direction, a tilted luminous visual frame, or more generally applying stimulation asymmetrically are directly expressed in imbalance of the body (cf. Wapner & Witkin, 1950; Witkin & Wapner, 1950).

The second level, *perceptual objectification*, is more complex insofar as it is characterized by a relationship between focal stimulation and organismic state that, in contrast to the first level of sensorimotor action, is differentiated. Here we see the entry of a principle described by Feibleman (1954) viz., each level is conceived of as integrating the prior level. On this second level, the internalized neuromuscular distribution characterizing the organismic state is now conceived of as a context to which focal environmental stimulation—a luminous rod—is related. We assume that perception, a reflection of the relationship between stimulation from the object and the organismic state, presupposes intentionality or directedness toward construction of an object. The differentiation between stimulation from an object and the organismic context makes such objectification possible. Objectification implies presence of a phenomenal object that is: (1) distanced (i.e., localized at a perceived distance from the body); (2) stabilized (i.e., experienced as permanent and possessing constant properties); and (3) articulated (i.e., differentiated into parts and properties that are hierarchically organized).

In treating the third level, *relations between perception of object out-there and perception of one's body* (two percepts reflecting operations occurring on the second level), these percepts are hierarchically integrated as components of the phenomenal field. Once the "body" and the "thing out-there" become differentiated aspects of experience, a variety of phenomenal relations between them are possible. For example, a body part may be experienced as an instrument of action directed outwardly as in pointing and/or touching an object or may be experienced as the passive object of an impinging force originating out-there (cf. Schlater, Baker, & Wapner, 1981).

The *development* of the relations among the different cognitive levels is exemplified in a study that utilizes the preceding theoretical and empirical paradigm. Specifically, the relation between things out-there (e.g., apparent verticality) and the self (e.g., perceived body position) was examined during ontogenesis (Wapner, 1968). It was found that, with increase in age, the position of apparent vertical shifts from the side of the body tilt to the side opposite body tilt and the apparent location of the longitudinal body axis deviates beyond true body tilt in all age groups with greater deviation with increase in age; for both apparent vertical and apparent body position, there is an accelerated increase after 13 years of age. This ontogenetic increase in angular disparity between perceived object and body positions is interpreted as a reflection of a shift toward greater differentiation between the body spatial and the external reference systems. This study provides a powerful methodology for studying the development of self-world relations or what Tobach (1987) has named "the category of self-non-self"; a critical dimension related to her analysis of integrative levels with respect to consciousness and externalization.

Using the same experimental situation, studies examining adaptation following disruption of perceptual objectification and experienced relations between percepts were conducted (Rierdan & Wapner, 1986, 1967; Werner & Wapner, 1955b). These studies were based on the following assumptions: (a) sensorimotor, perceptual (of body and of object), and conceptual aspects of cognition are alternative means of achieving knowledge about or "coming to terms with" the world; and (b) a crucial aspect of coming to terms with the world involves the establishment of a stable correspondence between body and object perception.

In these experiments, the correspondence between body and object that obtains under normal circumstances (the body is experienced as erect, and the object world is seen as erect) is disrupted with the introduction of rotating prisms (the body still feels erect, but the world is seen as tilted). Under these circumstances, an essential feature of adaptation is the establishment of a new correspondence between body perception and object perception.

It was found that changes in organism-in-environment relations, operative at the perceptual level in terms of a correspondence between body and object perception, are also operative at the conceptual level in terms of "self-world" attitudes. Just as self-world attitudes affect body and object perception as the individual structures his/her perceptual world, such attitudinal variations affect the relation between body and object perception as the individual *re*structures his/her world. If we manipulate these attitudes—one group is directed to attend to the body; the other group to objects-out-there—while introducing rotating prisms, then adaptation is expected to vary. It was found that opposite locations of apparent body position relative to apparent vertical occur after the introduction of prisms dependent on body-directed versus object-directed adaptation conditions; that is, under the object-directed condition, apparent body position is located increasingly counterclockwise of apparent verticality, whereas under the body-directed condition the opposite obtains.

Thus, this paradigm suggests that general analyses, developmental analyses, and analyses concerning adaptation following perturbation (disruption) of an organism's ongoing mode of integration represent a productive way of studying one aspect of the problem of levels of integration. However, this analysis is limited insofar as it is restricted to the psychological level. The critical problem remains of understanding how the biological, psychological, and sociocultural levels of integration are interrelated. What rules can we come up with to account for how a perturbation at one level of organization makes for changes at all other levels of organization?

Relations Among Biological, Psychological, and Sociocultural Levels of Integration. One approach to the study of relations among biological, psychological, and sociocultural levels is illustrated in our paradigmatic studies of critical person-in-environment transitions across the life-span (Wapner, 1977, 1981, 1987; Wapner & Demick, 1987a; Wapner, Kaplan & Cohen, 1973; Wapner, Ciottone, Hornstein, McNeil, & Pacheco, 1983). Investigations were initiated to study the interrelationships among cognition, affect, and valuation in the experience of, orientation in, and transactions with the everyday macroenvironment at even more complex levels than had been attempted in earlier work. More specifically, these studies assume that a potent perturbation to any part of the person-in-environment system at any level of integration (e.g., physical, psychological, and sociocultural aspects of the person, of the environment, or of both) impacts the system as a whole. We are, therefore, attempting to study critical transitions in a broad variety of situations where the perturbation has occurred and/or is induced in the environment, in the person, or in both.

For example, studies examining transitions initiated at the *physical level of the environment* have included: an assessment of the transactions of antisocial and schizophrenic patients prior to and following the relocation of a psychiatric therapeutic community (Demick & Wapner, 1980); an evaluation of the impact of urban renewal on the cognitive and affective experience of a city by individuals living in high- and low-socioeconomic neighborhoods (Demick, Hoffman, & Wapner, 1985); an assessment of the role of cherished possessions in adaptation of the elderly to a nursing home (Wapner, Demick, & Redondo, in press); and an analysis of alcoholics over a period of 5 weeks from entrance to time of leaving a hospital setting (Demick & Peicott, 1984; Demick, Peicott, & Wapner, 1985; Demick & Wapner, 1986). With respect to the *interpersonal level of the environment* (i.e., other people constituting the environment), studies have included the development of friendships (Thomason, 1986) and the development of personal networks following entrance into college (Minami, 1986). At the *sociocultural level of the environment*, studies have included: entering a nursery school (e.g., Ciottone & Quirk, 1985; Quirk, Ciottone, Minami, Wapner, Yamamoto, Ishii, Lucca-Irizarry, & Pacheco, 1986; Quirk, Sexton, Ciottone, Minami, & Wapner, 1984); entering and adapting to medical education (Quirk, Ciottone, Letendre, & Wapner, 1987); migration and return migration of adolescents between Puerto Rico and the U.S. Mainland (Pacheco, Lucca, & Wapner, 1985); and the impact of mandatory legislation on use of automobile seat belts in the United States and Japan (Wapner, Demick, Inoue, Ishii, & Yamamoto, 1986; Demick, Inoue, Wapner, Ishii, Minami, Yamamoto, & Nishiyama, 1986).

Further, studies conducted on critical transitions following change in the status of the *person* have included: on the *biological* level, the onset of diabetes (Collazo, 1985); on the *psychological* level, the onset of various psychopathologies (Demick, 1986); and on the *sociocultural* level, transition to parenthood (Clark, 1987) and to retirement (Hornstein & Wapner, 1984, 1985; Wapner & Demick, 1986; Wapner, Demick & Damrad, 1988).

To illustrate the ways in which the concept of levels of integration enter into the analyses of critical transitions, we consider two examples. In the first example, the perturbation is introduced to the *person* aspect of the person-in-environment system at the least complex (i.e., biological, level). In the second, the perturbation is introduced at the most complex, *environmental* aspect of the person-in-environment system (viz., at the sociocultural level).

First is the work in progress of Collazo (1985), who is studying the transition from health to illness as exemplified in the onset of diabetes. More specifically, the study is concerned with analyzing a number of relations between the focal person (self) and other parts of the system, including: (a) relations between one's psychological self (e.g., self-esteem) and one's biological self (e.g., unstable glucose levels); (b) transactions of self in relation to the physical aspect of the environment (e.g., some diabetics are unwilling to move beyond the restricted physical area of their home community having deep concern for availability of insulin supplies, etc.) as well as to the interpersonal aspect of the environment (e.g., some diabetics wonder whether they should ever get married); and (c) transactions related to the sociocultural context (e.g., changes in values and behaviors of the diabetic related to culturally defined attitudes toward the sick individual).

The second example comes from a series of related studies concerning experience and action with respect to automobile seat belt usage (see Wapner & Demick, 1987c, for a review of these studies). Semi-independent studies were conducted prior to and following the initiation of mandatory seat belt legislation (a perturbation at the sociocultural level of the environment) in the United States (Massachusetts) and Japan (Hiroshima). Comparison of the findings indicated general cultural differences as well as cultural differences over time.

For example, results provided support for cultural similarities and differences related to: (a) general factors that predispose individuals to the general action/nonaction of using seat belts; (b) precursors activating the specific behavior of "buckling up" when one enters the automobile; and (c) action (e.g., Japanese wear seat belts on the highway more often than Americans) and experience of action (e.g., relative to Americans, Japanese feel "virtuous" but not "confident" when wearing seat belts). As regards cultural differences over time, the usage rates began to level off and remained constant or increase further in Japan; however, they continued to decrease steadily in the United States.

In addition to highlighting the need to complement the *cognitive* aspects of automobile seat belt usage with *affective* and *valuative* aspects, these studies have implications from a holistic perspective for the larger problems of relations between individual experience and action and types of relations among levels.

Experience and action. To handle some aspects of the complexity of the relationships between experience and action, we have found it useful to employ Turvey's (1977) musical instrument metaphor in our analysis. This metaphor distinguishes between "tuning" and "activating" inputs. As Gallistel (1980) succinctly stated: "Consider for example a piano or guitar. Turning the tuning pegs to tighten or loosen the strings does not produce music, but it profoundly alters the music that is produced. The tightness of each string is a parameter of a piano or guitar. Signals that adjust these parameters are called parameter adjusting or tuning inputs to distinguish them from activating inputs like key-strikes and string-plucking" (p. 364).

More specifically, in the analysis just mentioned, tuning corresponds to the category of *general factors* preparing the individual for the action/nonaction of seat belt usage (e.g., knowledge of seat belt effectiveness; generalized anxiety about driving; prior experience in accident). We have found that such factors are necessary but not sufficient for usage/nonusage. *Specific precursors* for triggering, initiating, and activating the concrete behavior of "buckling up" in the particular context of the automobile (e.g., imagination of accidents, feeling "undressed," desire to serve as a role model for children) are analogous to striking a key, which directly leads to the action of seat belt usage, provided the general factors have been met.

Types of Relations Among Levels. Can such metaphors as Turvey's be used to attack a related problem of potential significance, namely, how experience and action at the individual level are related to experience and action at the sociocultural level? Various complexities enter into an analysis designed to answer this question. For example, the goals and instrumentalities governing action at the sociocultural level differ from the goals and instrumentalities governing action at the individual or interpersonal levels. Instituting a change in environmental policy at the sociocultural level depends on some form of group consensus process, whereas instituting a change such as the shift from Non-user to Variable-user, or Variable-user to Committed-user of seat belts, depends on an individual psychological process, namely, perception, or, more generally, experience in relation to action.

Previous work on the relations among cognitive processes has also suggested ways in which one might conceptualize the relations between individual action and societal rules and regulations. For example, the relationships between the individual and society can be *supportive* (e.g., personal experience—such as being saved from injury by a seat belt—makes one an advocate for mandatory legislation), *antagonistic* (e.g., personal experience—such as a relative being hurt in an accident by wearing seat belts—and public policy requiring seat belt use are antithetical), *substitutive/vicarious* (e.g., individuals who vote in favor of mandatory legislation, substituting the good of society for their own individual feelings of loss of personal freedom), or *correspondent* (e.g., personal opinion and public policy coincide) between these two levels. Can these modes of conceptualization be of help in understanding relations between experience and action? Can they help in tailoring modes of intervention that can change individual action so that it is in conformity with individual goals? Can they throw light on the distinction between individual compliance and commitment?

THE FUTURE

In summary, although we have made inroads into specifying some of the relations both between and within levels of integration, the problem remains of further defining the nature of the conditions that makes for one or the other of these relationships and their development broadly conceived to include ontogenesis, pathogenesis, and microgenesis. These are the goals of our ongoing research programs, which shall continue to utilize such paradigms as those treating critical person-in-environment transitions as well as to examine more directly, through experimental manipulations, these and other developmental relations among levels of integration in human functioning.

REFERENCES

Allport, F. H. (1955). *Theories of perception and the concept of structure*. New York: Wiley.

Anthony, E. J. (1957). The system makers: Piaget and Freud. Symposium on the contribution of current theories to an understanding of child development. *British Journal of Medical Psychology, 30*, 255–269.

Aronson, L. R. (1987). Some remarks on integrative levels. In G. Greenberg & E. Tobach (Eds.), *Cognition, language and consciousness: Integrative levels. The T. C. Schneirla Conference Series* (Vol. 2). Hillsdale, NJ: Lawrence Erlbaum Associates.

Aronson, L. R., Tobach, E., Lehrman, D. S., & Rosenblatt, J. S. (Eds.). (1972). *Selected writings of T. C. Schneirla*. San Francisco: Freeman.

Ayala, F. J. (1970). Teleological explanations in evolutionary biology. *Philosophy of Science, 37*, 1–15.

Baker, A. H., Rierdan, J., & Wapner, S. (1974). Age changes in size-value phenomena. *Child Development, 45*, 257–268.

Blake, R. R., & Ramsey, G. V. (1951). *Perception: An approach to personality*. New York: Ronald.

Bruner, J. S., & Goodman, C. C. (1947). Value and need as organizing factors in perception. *Journal of Abnormal and Social Psychology, 42*, 33–44.

Campbell, D. T. (1974). 'Downward causation' in hierarchically organised biological systems. In F. J. Ayala & T. Dobzhansky (Eds.), *Studies in the philosophy of biology*. London: Macmillan.

Ciottone, R., & Quirk, M. (1985, March). *The integration of two worlds: Home and preschool*. Paper presented at Eastern Psychological Association Annual Meeting, Boston.

Clark, W. (1987). *Transition to parenthood*. Doctoral dissertation in preparation. Worcester, MA: Clark University.

Collazo, J. A. (1985, March). *Transition from health to illness: Experiential changes following the onset of diabetes*. Presented at a Symposium on 'Environmental Psychology: In Search of a Theory' at the Eastern Psychological Association Annual Meeting, Boston.

Decarie, T. G. (1963). *Intelligence and affectivity in early childhood* (Translated by E. P. Brandt & L. W. Brandt). New York: International Universities Press.

Demick, J. (1986). *Transition to psychopathology*. Manuscript in preparation.

Demick, J., Hoffman, A., & Wapner, S. (1985). Residential context and environmental change as determinants of urban experience. *Children's Environments Quarterly, 2*(3), 44–54.

Demick, J., Inoue, W., Wapner, S., Ishii, S., Minami, H., Yamamoto, T., & Nishiyama, S. (1986, August). *Experience and action of automobile seat belt usage: United States and Japan*. Paper presented at the American Psychological Association Annual Meeting, Washington, DC.

Demick, J., & Peicott, J. (1984, August). *Temporal changes in hospitalized alcoholics' cognitive organization of space*. Paper presented at American Psychological Association Annual Meeting, Toronto, Canada.

Demick, J., Peicott, J., & Wapner, S. (1985, August). *Temporal changes in hospitalized alcoholics' self-world relationships*. Paper presented at American Psychological Association Annual Meeting, Los Angeles.

Demick, J., & Wapner, S. (1980). Effect of environmental relocation on members of a psychiatric therapeutic community. *Journal of Abnormal Psychology, 89*, 444–452.

Demick, J., & Wapner, S. (1986, July). *Entering, living in, and leaving the psychiatric setting*. Paper presented at the 21st Congress of the International Association of Applied Psychology, Jerusalem, Israel.

Dewey, J., & Bentley, A. F. (1949). *Knowing and the known*. Boston: Beacon.

Dowling, R. M. (1965). Visual recognition threshold and concurrent motor activity. *Perceptual and Motor Skills, 20*, 1141–1146.

Dowling, R. M., Werner, H., & Wapner, S. (1960, April). *Effect of motor activity on visual recognition threshold*. Paper presented at Eastern Psychological Association Annual Meeting, New York.

Feibleman, J. K. (1954). Theory of integrative levels. *British Journal of Philosophy of Science, 5*, 59–66.

Fraenkel, G. S., & Gunn, D. L. (1961). *The orientation of animals: Kinesis, taxes and compass reactions*. (reprint of original 1940 edition). New York: Dover Publications.

Franklin, M. B. (1981). Perspectives on theory: Another look at the developmental-interaction point of view. In E. K. Shapiro & E. Weber (Eds.), *Cognitive and affective growth: Developmental interaction*. Hillsdale, NJ: Lawrence Erlbaum Associates.

Freimuth, M., & Wapner, S. (1979). The influence of lateral organization on the evaluation of paintings. *British Journal of Psychology, 70*, 211–218.

Gallistel, C. R. (1980). *The organization of action: A new synthesis*. Hillsdale, NJ: Lawrence Erlbaum Associates.

Goldman, A. E. (1953). Studies in vicariousness: Degree of motor activity and the autokinetic phenomenon. *American Journal of Psychology, 66*, 613–617.

Greenberg, G., & Tobach, E. (1984). *Behavioral evolution and integrative levels. The T. C. Schneirla Conference Series*. Hillsdale, NJ: Lawrence Erlbaum Associates.

Herrick, C. J. (1949). A biological survey of integrative levels. In R. W. Sellars, V. J. McGill, & M. Farber (Eds.), *Philosophy for the future*. New York: Macmillan.

Hofstadter, A. (1941). Objective teleology. *Journal of Philosophy, 38*, 29–39.

Hornstein, G. A., & Wapner, S. (1984). The experience of the retiree's social network during the transition to retirement. In C. M. Aanstoos (Ed.), *Exploring the lived world readings in phenomenological psychology*. Carrollton, GA: West Georgia College Press.

Hornstein, G. A., & Wapner, S. (1985). Modes of experiencing and adapting to retirement. *International Journal on Aging and Human Development, 21*(4), 291–315.

Isaac, D. M. (1958). *The effect of interpersonal feelings on psychological distance*. Unpublished master's thesis. Clark University, Worcester, MA.

Kaplan, B. (1966). The study of language in psychiatry. In S. Arieti (Ed.), *American handbook of psychiatry* (Vol. III). New York: Basic Books.

Kaplan, B. (1967). Meditations on genesis. *Human Development, 10*, 65–87.

Kaplan, B. (1983). Trio of trials. In R. Lerner (Ed.), *Developmental psychology: Historical and philosophical aspects*. Hillsdale, NJ: Lawrence Erlbaum Associates.

Kaplan, B., Wapner, S., & Cohen, S. B. (1976). Exploratory applications of the organismic-developmental approach to man-in-environment transactions. In S. Wapner, S. B. Cohen, & B. Kaplan (Eds.), *Experiencing the environment*. New York: Plenum Press.

Krus, D. M., Resnick, O., & Raskin, M. (1966). Apparent eye level test: Its background and use in psychopharmacology. *Archives of General Psychiatry, 14*, 419–427.

Krus, D. M., Werner, H., & Wapner, S. (1953). Studies in vicariousness: Motor activity and perceived movement. *American Journal of Psychology, 66*, 603–608.

Langer, J., Wapner, S., & Werner, H. (1961). The effect of danger upon the experience of time. *American Journal of Psychology, 74*, 94–97.

Langer, J., Werner, H., & Wapner, S. (1965). Apparent speed of walking under conditions of danger. *Journal of General Psychology, 73*, 291–298.

Lavine, T. (1950). Knowledge as interpretation: An historical survey. *Philosophy and Phenomenological Research, 10*, 526–540; *11*, 80–103.

Lazarus, R. S. (1981). A cognitivist's reply to Zajonc on emotions and cognition. *American Psychologist, 36*, 222–223.

Lazarus, R. S. (1982). Thoughts on the relations between emotions and cognition. *American Psychologist, 37*, 1019–1024.

Miller, A. (1963). Verbal satiation and the role of concurrent activity. *Journal of Abnormal and Social Psychology, 66*, 206–212.

Minami, H. (1986). Establishment and transformation of personal networks during the first year of college: A developmental analysis. *Dissertation Abstracts International*, DA8608772.

Novikoff, A. B. (1945a). The concept of integrative levels and biology. *Science, 101*, 209–215.

Novikoff, A. B. (1945b). Continuity and discontinuity in evolution. *Science, 102*, 405–406.

Pacheco, A. M., Lucca, N., & Wapner, S. (1985). The assessment of interpersonal relations among Puerto Rican migrant adolescents. In R. Diaz-Guerro (Ed.), *Cross-cultural and national studies in social psychology*. North Holland: Elsevier Science.

Piaget, J. (1952). Autobiography. In E. G. Boring, H. S. Langfeld, R. M. Yerkes, & H. Werner (Eds.), *A history of psychology in autobiography*, Worcester, MA: Clark University Press.

Piaget, J. (1960). The general problems of the psychological development of the child. In J. M. Tanner & B. Inhelder (Eds.), *Discussions on child development, Vol. IV: The proceedings of the fourth meeting of the World Health Organization study group on the psychological development of the child*. New York: International Universities Press.

Piaget, J. (1971). *Insight and illusions of philosophy* (translated by Wolfe Mays). New York: World.

Quirk, M., Ciottone, R., Letendre, D., & Wapner, S. (1987). Critical person-in-environment transitions in medical education. *Medical Teacher, 9* (No. 4), 415–423.

Quirk, M., Ciottone, R., Minami, H., Wapner, S., Yamamoto, T., Ishii, S., Lucca-Irizarry, N., & Pacheco, A. (1986). Values mothers hold for handicapped and non-handicapped preschool children in Japan, Puerto Rico, and the United States Mainland. *International Journal of Psychology, 21*, 463–485.

Quirk, M., Sexton, M., Ciottone, R., Minami, H., & Wapner, S. (1984). Values mothers hold for handicapped and non-handicapped preschoolers. *Merrill-Palmer Quarterly, 30*(4), 403–418.

Rierdan, J., & Wapner, S. (1966). Experimental study of adaptation to visual rearrangement deriving from an organismic-developmental approach to cognition. *Perceptual and Motor Skills, 23*, 903–916.

Rierdan, J., & Wapner, S. (1967). Adaptive changes in the relationship between visual and tactual-kinesthetic perception. *Psychonomic Science, 7*, 61–62.

Rosenblatt, B. P. (1956). The influence of affective states upon the body image and upon the perceptual organization of external space. *Dissertation Abstracts International, 16/09*, 17–21. (University Microfilms No. 18088)

Schlater, J. A., Baker, A. H., & Wapner, S. (1981). Apparent arm length with active versus passive touch. *Bulletin of the Psychonomic Society, 18*, 151–154.

Schneirla, T. C. (1949). Levels in the psychological capacities of animals. In R. W. Sellars, V. J. McGill, & M. Farber (Eds.), *Philosophy for the future*. New York: Macmillan.

Schneirla, T. C. (1957). The concept of development in comparative psychology. In D. B. Harris (Eds.), *The concept of development*. Minneapolis: University of Minnesota Press.

Schneirla, T. C. (1972). Modifiability in insect behavior. In L. R. Anderson, E. Tobach, J. S. Rosenblatt, & D. S. Lehrman (Eds.), *Selected writings of T. C. Schneirla*. San Francisco: Freeman.

Switzer, J. (1961). Developmental differences in place and name sequence learning in normal, hyperactive, and hypoactive eight- and twelve-year-old boys. *Dissertation Abstracts International, 22/07*, 2482 (University Microfilms No. 61-5013)

Teft, L., Wapner, S., Werner, H., & McFarland, J. H. (1964). Relation between perceptual and conceptual operations: Numerical distance and visual extent. *British Journal of Psychology, 55*, 421–427.

Thomason, D. (1986, April). *Growing friendships: A developmental analysis*. Presented at the Annual Meeting of the Eastern Psychological Association, New York.

Thompson, N. S. (1987). Ethology and the birth of comparative teleonomy. In R. Campan & R. Zayan (Eds.), *Relevance of models and theories in ethology*. Toulouse, France: Privat I. E. C.

Tobach, E. (1987). Integrative levels in the comparative psychology of cognition, language, and consciousness. In G. Greenberg & E. Tobach (Eds.), *Cognition, language and consciousness: Integrative levels. The T. C. Schneirla Conference Series* (Vol. 2). Hillsdale, NJ: Lawrence Erlbaum Associates.

Turvey, M. T. (1977). Preliminaries to a theory of action with reference to vision. In R. Shaw & J. Bransford (Eds.), *Perceiving, acting, and knowing*. Hillsdale, NJ: Lawrence Erlbaum Associates.

von Bertalanffy, L. (1950). The theory of open systems in physics and biology. *Science, III*, 23–29.

Wapner, S. (1968). Age changes in perception of verticality and of the longitudinal body axis under body tilt. *Journal of Experimental Child Psychology, 6*, 543–555.

Wapner, S. (1969). Organismic-developmental theory: Some applications to cognition. In Langer, Mussen, & Covington (Eds.), *Trends and issues in developmental psychology*. New York: Holt, Rinehart, & Winston.

Wapner, S. (1977). Environmental transition: A research paradigm deriving from the organismic-developmental systems approach. In L. van Ryzin (Ed.), *Wisconsin conference on research methods in behavior environment studies proceedings*. Madison: University of Wisconsin.

Wapner, S. (1978). On Professor Jean Piaget's eightieth birthday. *Proceedings of the XXIst International Congress of Psychology*. Paris: Presses Universitaires de France.

Wapner, S. (1981). Transactions of persons-in-environments: Some critical transitions. *Journal of Environmental Psychology, 1*, 223–239.

Wapner, S. (1983). Living with radical disruption of person-in-environment systems. *IATSS Review, 9*(2), June, 133–148. (In Japanese. English translation available.)

Wapner, S. (1986, November). *An organismic-developmental systems approach to the analysis of experience and action*. Presented at a Conference on 'Holistic Approaches to Analysis of Experience and Action' at the University of Catania, Sicily.

Wapner, S. (1987). A holistic, developmental, systems-oriented environmental psychology: Some beginnings. In D. Stokols & I. Altman (Eds.), *Handbook of environmental psychology*. New York: Wiley.

Wapner, S., Ciottone, R., Hornstein, G., McNeil, O., & Pacheco, A. M. (1983). Critical transitions through the life cycle. In S. Wapner & B. Kaplan (Eds.), *Toward a holistic developmental psychology*. Hillsdale, NJ: Lawrence Erlbaum Associates.

Wapner, S., Cirillo, L., & Baker, A. H. (1969). Sensory-tonic theory: Toward a reformulation. *Archivio Di Psicologia Neurologia E. Psichiatria, XXX*, 493–512.

Wapner, S., Cirillo, L., & Baker, A. H. (1971). Some aspects of the development of space perception. In J. P. Hill (Ed.), *Minnesota symposium on child psychology*. (Vol. V). Minneapolis: University of Minnesota Press.

Wapner, S., Cohen, S. B., & Kaplan, B. (Eds.). (1976). *Experiencing the environment*. New York: Plenum Press.

Wapner, S., & Demick, J. (1986). *Retirement study. Working paper*. Worcester, MA: Clark University.

Wapner, S., & Demick, J. (1987a, August). *A holistic, developmental, systems approach to person-in-environment functioning*. Presented as part of the Symposium 'Unifying the Science: Strategies from Psychology's Grand Theories' at the American Psychological Association Annual Meeting, New York.

Wapner, S., & Demick, J. (1987b). *Some relations among cognition, emotion, and action: A holistic, developmental, systems approach*. Unpublished manuscript.

Wapner, S., & Demick, J. (1987c, August). *A process-orientated analysis of risk: Automobile seat belt usage*. Presented as part of symposium "Public Perception of Technological and Natural Hazards" at the American Psychological Association Annual Meeting, New York.

Wapner, S., Demick, J., & Damrad, R. (1988, April). *Transition to retirement: Eight years after*. Presented at Eastern Psychological Association Annual Meeting, Buffalo, NY.

Wapner, S., Demick, J., Inoue, W., Ishii, S., & Yamamoto, T. (1986). Relations between experience and action: Automobile seat belt usage in Japan and the United States. In W. H. Ittelson, M. Asai, & M. Carr (Eds.), *Proceedings of the 2nd USA/Japan seminar on environment and behavior*. Tucson, AZ: Department of Psychology, U. of Arizona.

Wapner, S., Demick, J., & Redondo, J. P. (in press). *Cherished possessions and adaptation of the elderly to the nursing home. International Journal of Aging and Human Development.*

Wapner, S., & Kaplan, B. (Eds.) (1983). *Toward a holistic developmental psychology*. Hillsdale, NJ: Lawrence Erlbaum Associates.

Wapner, S., Kaplan, B., & Ciottone, R. (1981). Self-world relationships in critical environmental transitions: Childhood and beyond. In L. Liben, A. Patterson, & N. Newcombe (Eds.), *Spatial representation and behavior across the life span*. New York: Academic Press.

Wapner, S., Kaplan, B., & Cohen, S. (1973). An organismic-developmental perspective for understanding transactions of men in environments. *Environment and Behavior, 5*, 255–289.

Wapner, S., & Rand, G. (1967). Age changes in verbal satiation. *Psychonomic Science*, *9*, 93–94.

Wapner, S., & Werner, H. (1957). *Perceptual development*. Worcester, MA: Clark University Press.

Wapner, S., & Werner, H. (1965). An experimental approach to body perception from the organismic-developmental point of view. In S. Wapner & H. Werner (Eds.), *The body percept*. New York: Random House.

Wapner, S., Werner, H., & Krus, D. M. (1957). The effect of success and failure on space localization. *Journal of Personality, 25*, 752–756.

Wapner, S., & Witkin, H. A. (1950). Visual factors in the maintenance of body balance. *American Journal of Psychology, 63*, 385–408.

Weisenburg, T., & McBride, K. E. (1935). *Aphasia: A clinical and psychological study*. New York: Commonwealth Fund.

Werner, H. (1957a). *Comparative psychology of mental development*. New York: International Universities Press. (Originally published in German, 1926, in English, 1940.)

Werner, H. (1957b). The concept of development from a comparative and organismic point of view. In D. Harris (Ed.), *The concept of development*. Minneapolis: University of Minnesota Press.

Werner, H., & Kaplan, B. (1956). The developmental approach to cognition: Its relevance to the psychological interpretation of anthropological and ethnolinguistic data. *American Anthropologist, 58*, 866–880.

Werner, H., & Kaplan, B. (1963). *Symbol formation*. New York: Wiley.

Werner, H., & Wapner, S. (1949). Sensory-tonic field theory of perception. *Journal of Personality, 18*, 88–107.

Werner, H., & Wapner, S. (1952). Toward a general theory of perception. *Psychological Review, 59*, 324–338.

Werner, H., & Wapner, S. (1955a). Changes in psychological distance under conditions of danger. *Journal of Personality, 24*, 153–167.

Werner, H., & Wapner, S. (1955b). The Innsbruck studies on distorted visual field in relation to an organismic theory of perception. *Psychological Review, 62*, 130–138.

Witkin, H. A., & Wapner, S. (1950). Visual factors in the maintenance of upright posture. *American Journal of Psychology, 63*, 21–50.

Wolff, P. H. (1960). The developmental psychologies of Jean Piaget and psychoanalysis. *Psychological Issues, 2*, 131.

Zajonc, R. B. (1980). Feeling and thinking: Preferences need no inferences. *American Psychologist, 35*, 151–175.

Zaporozhets, A. V. (1965). The development of perception in the preschool child. In P. H. Mussen (Ed.), *European research in cognitive development. Monograph of the Society for Research in Child Development, 30* (2, Serial No. 100).

Zener, K. (Ed.). (1949a). Interrelations between perception and personality: A symposium (special issue). *Journal of Personality, 18*(1).

Zener, K. (Ed.) (1949b). Interrelations between perception and personality: A symposium (special issue). *Journal of Personality, 18*(2).

5 Developmental Epistemology and Self-Knowledge: Towards a Reinterpretation of Self-Esteem

Robert B. Cairns
University of North Carolina at Chapel Hill

The dichotomy between *mind and behavior* has proved to be as beguiling and problematic in comparative psychology as have two other primary dichotomies: *nature* and *nurture*, and *ontogeny* and *phylogeny*. It is perhaps inevitable that close examination of one of the dichotomies should call up the issues associated with the other two. To adumbrate the conclusions of this chapter, it seems indisputable that the biological design of the information-processing system places constraints on what information is received and how it is then constructed and elaborated. Evolution assures variations in neural input and information integration that are serviceable for the species. For a developmentalist, however, that is only part of the story. A comparable process of epistemological constraint and construction recurs in the ontogeny of each organism in ways that ensure its proper functioning and adaptation.

What I have to say about the development of knowledge and its limits may appear at first blush to be tangential to—or contradictory with—evolutionary epistemology.[1] If that impression persists, then this chapter will have failed. Rather than supplanting naturalistic epistemology, a balanced view of ontogeny can extend an evolutionary theory of knowledge and correct some of its shortcomings. To date, developmental views—particularly those of T. C. Schneirla (1966) and Z-Y. Kuo (1967)—have been given short shrift in otherwise masterful efforts to address the issues of evolutionary epistemology (e.g., Campbell, 1974;

[1]In this chapter, the focus is on a particular domain of knowledge, namely, knowledge about the self and, secondarily, of other persons. It would be informative to attempt to extend the argument to knowledge about the physical world and external reality.

Lorenz, 1962). Inattention to development may deprive the evolutionary perspective of a rich source of information about how individuals actually gain and organize knowledge in their lifetimes. I argue in this chapter that modified versions of evolutionary epistemology and developmental epistemology can prove to be mutually dependent and reciprocally supportive. Taken together, they may provide useful guides for understanding how individuals and species organize information about their worlds. They also lead to a reconceptualization of the role of cognition and self-esteem in social development.

EVOLUTION, EPISTEMOLOGY, AND THE HOLISTIC PERSPECTIVE

In one of his final scientific papers, T. C. Schneirla (1966) provided an agenda for the discipline and specified what were the hazards of ignoring developmental issues. He wrote: "Behavioral ontogenesis is the backbone of comparative psychology. Shortcomings in its study inevitably handicap other lines of investigation from behavioral evolution and psychogenetics to the study of individual and group behavior" (p. 280). Note that there is no mention of the study of knowledge or consciousness. The omission is perhaps not surprising, given that sensation, perception, and information processing were supposed to be subsumed by the concept of "behavior" (e.g., cognitive behavior). But if the holistic model is ever to effectively address distinctive issues of the human condition, it must explicitly consider the role of cognition and consciousness.

There is a problem of intellectual territoriality. The turf of the development of consciousness was originally claimed by the "genetic epistemology" of Jean Piaget, Heinz Werner, and James Mark Baldwin. Any new claims on the matter must necessarily be compared to and contrasted with the earlier views of cognitive developmental writers. In the past, the battles of psychobiology have been fought with classical ethology on the right and learning theory and behaviorism on the left. On this score, Gerard Piel (1970) stated the matter succinctly when he wrote: "The animal of Schneirla's studies is . . . a productively different kind of animal from the ethologist's wind-up toy, awaiting its releaser, or from the operant-behaviorist's feathered and furry computers, so susceptible to programming and read-out" (p. 6).

After two decades, those battles about animal behavior do not seem as interesting as they once were. Learning theory—as it was originally developed by Thorndike, Hull, Guthrie, and Skinner—can no longer claim hegemony in the science. When it became obvious that one cannot ignore (a) the characteristics of the species, (b) the nature of the task, and (c) the context in which learning occurs, the elaborate models of learning in vacuo faded from center stage. Ethological theory has undergone an equally sharp reformation, though it has not diminished in importance. Many of the major developmental themes of Schneirla and Kuo have now become incorporated into the agenda of modern ethology.

It is quite another matter with genetic epistemology. The term and concerns were introduced by James Mark Baldwin (1915), and they have both since become identified with Jean Piaget. Baldwin and Piaget both addressed the development of mind and consciousness as a first priority. Together with Heinz Werner and modern expositors (e.g., Kohlberg, 1969), these theorists have led the way toward an interpretation of cognition from a developmental perspective. The mind of the child is qualitatively different from the adult in content as well as in its structures and processes. These assumptions on the ontogeny of thinking led immediately to a revised perspective on knowledge and what is known. Whereas Piaget borrowed the terminology and concepts of biology, his concerns were with the development of language and thought in children. In other words, he focused on the software of knowledge. He did not address the development of the perceptual equipment, the neurological pathways, and the mechanisms of biobehavioral integration. This hardware of biology and behavior was the domain of holistic theory; namely, the developmental psychobiology of Schneirla (1966), Kuo (1967), and their colleagues.

Genetic Epistemology

It should be useful at this juncture to summarize some of the guiding assumptions of the body of theory that has been subsumed by the term, genetic epistemology. The label and the ideas were introduced by J. M. Baldwin (1915) and subsequently extended by Piaget (1926, 1952), Werner (1940), and Kohlberg (1969). Any attempt to briefly describe a set of concepts as complex as that of genetic epistemology is bound to be inadequate or misleading.[2] Four assumptions from the Piagetian–Baldwinian view of cognitive development seem especially relevant of the current discussion (from Cairns, 1979):

1. Cognitive development follows principles that are analogous to those of general biological development.
2. Intelligence requires distancing oneself from the "here and now," the events in the immediate present. Hence cognitive development can be described in terms of successive stages of decentering one's mental conceptions of events from the immediate sensations and perceptions. This leads to abstract representations of those events.
3. The cognitive structures of the child ("schemata") undergo progressive and invariant sequential organization during development. Cognitive develop-

[2]See Flavell (1977) and Piaget (1926, 1952) for presentations of these issues.

ment thus involves an integration of earlier cognitive structures into a hierarchy of increasing complexity (simple undifferentiated structures precede complex, differentiated ones.)

4. True novelties arise during the course of development that are irreversible and cannot be reduced to earlier stages of development.

Both Baldwin and Piaget employ biological principles, but genetic epistemology is not a biological theory of cognitive development. Unlike ethological or psychobiological models of perception and cognition, genetic epistemology has not been concerned with the neural substrate of intelligence, nor with evolutionary causation. It is an analogue to biological development because cognitive structures of the child ("schemata") are assumed to undergo reorganization in ontogeny. Cognitive changes in development are stimulated by both internal and external events. The changes permit children to achieve an equilibrium between their existing mental schemata and variations in the external world.

Development thus provides the basis for changes in the views that children have of themselves and the world around them. The reorganization follows a predictable, invariant, and irreversible sequence. Children graduate in their thinking in stages, beginning with a dependence on immediate sensation in infancy and continuing to conceptual independence and abstraction in adolescence. Earlier schemata are integrated into emergent ones, and it is through this hierarchical integration that the child's mental representations become independent of the constraints of time and place. The hierarchical integration of stages—global, undifferentiated stages precede differentiated, integrated ones—has been called the *orthogenetic principle* (Werner, 1940; also K. E. von Baer, 1828).

Applying a correlated set of principles to the linkages between social behavior and cognitive development, Lawrence Kohlberg (1969) argued for cognitive primacy in social behavior. Social cognitions set the stage for all forms of social patterns, from sex typing and modeling (imitation) to aggression and attachment. The child's understanding of the social world, and his/her self-categorization with respect to it, provides the basis for action. In this respect, Kohlberg (1969) was at the forefront in reviving J. M. Baldwin's (1897) concept of the self as an organizing, active agent in socialization. Developmental changes in cognition and the child's understanding of the beliefs and attitudes of others go hand in hand with changes in personality and behavior.

On this score, Baldwin (1897) wrote: "The development of the child's personality could not go on at all without the constant modification of his sense of himself by suggestions from others. So he himself, at every stage, is really in part some one else, even in his own thought of himself." By integrating developmental changes in cognitive representation with changes in information from the social context, Baldwin depicts persons and their knowledge about themselves as being a unique product of maturation and society. The matter is not only epistemological, a question of knowledge; it is a question of reality and the changing personal standards by which reality is assessed.

According to Baldwin (1915), the level and type of social organization in society reflects the level of the individual's cognitive development:

> If the development of the apprehension of self and things, in the individual's case, is correctly made out, we may be able to trace out the correlative stages in the historical evolution of the race. The social products could not rise higher, at any period, than the individuals' knowledge and practice would justify—so much negatively. And the notable and typical gains of individuals, in mental and moral progress, would show themselves, in time and in serial order, in the texture of the social fabric; in its institutions and in its theoretical speculations—so much, at least, more positively. (Vol. 4, p. 6)

In the same vein, Kohlberg (1969) argued that the more cognitively advanced the species, the more social it will be. On this model, the level of cognitive functioning jointly determines individual adaptations and the adaptations of the society and species.

Holistic Theory: Principles and Distinctions

Is there anything fresh in the Schneirla–Kuo perspective on organismic development that has not already been explored in the Baldwin–Piaget models of genetic epistemology? To address that question, it is first appropriate to define the holistic orientation in terms of its methods and constructs.[3] These characteristics can be summarized in two metatheoretical guidelines and six developmental principles.

One of the metatheoretical guidelines concerns that need for parsimony in explanation. For example, the problem that Schneirla (1966) found with instinct was *not* in the assumption that there are robust, inherent constraints on behavior. To the contrary, Schneirla and those inspired by him were thoroughly "biological" in their research and their concepts. For these researchers, a major problem with the concept of instinct as employed by Lorenz (1965) was that multiple drives and dispositions were required to explain phenomena that might be more efficiently described in terms of quantitative biological principles. According to Schneirla (1966) and Kuo (1967), the concept of instinct provided answers when more questions were called for. As Lehrman (1970) observed, much of this debate reflected differences in the level of analysis preferred by theorists, whether proximal (ontogenetic) or ultimate (phyletic).

---

[3] "Orientations" seems more appropriate than "theory," because the major contributors to the holistic approach to developmental psychobiology have not presented a cohesive theoretical model so much as a set of principles to guide the analysis of developmental phenomena.

A second metatheoretical bias concerns the preference for induction in scientific research and explanation. As Quine (1981) has observed: "a good scientific theory is under tension from two opposing forces: the drive for evidence and the drive for system" (p. 31). For those who followed Schneirla's lead, research begins with an analysis of a behavior's development, its controls, and its consequences. Explanatory statements on origins and determinants should be anchored first in a network of empirical relationships. This view can be contrasted with the psychological tradition of formal hypothesis testing, where much research seems designed to *demonstrate* principles rather than to *discover* them. In Quine's terms, the drive for evidence is at least as strong as the drive for system. In contrast, "genetic epistemology" and associated theories of cognitive development have from the beginning given the priority to systemization and theory. Once they were formulated, models of genetic epistemology from Baldwin through Piaget and Kohlberg have relied on demonstrational evidence.[4] On this point—reliance on empirical data and experimental analyses in theory building—scientists in the holistic tradition differed markedly from those concerned with the study of genetic epistemology. This difference proved to be decisive.

Beyond these metatheoretical biases, the principles that have guided research in the holistic tradition of Schneirla and Kuo seem remarkably similar to the assumptions adopted by Baldwin and Piaget. In brief, six relevant principles of holistic theory may be summarized (from Cairns, 1979):

1. Behavior is appropriately viewed in terms of an organized biological system, and its explanation requires holistic analyses. For example, in addressing the nature–nurture issue, Schneirla (1966) wrote: "The developmental contributions of the two complexes, maturation and experience, must be viewed as fused (i.e., as inseparably coalesced) at all stages in the ontogenesis of any organism" (p. 288).
2. To understand behavioral development, information must be obtained about the full context of ontogenesis, including the social ecology in which development occurs and the reciprocal relationships in which individuals become embedded. This "holistic" analysis across time is required because bidirectional feedback effects can occur with the organism and across organisms.
3. There is a continuity in development, such that the organization at one stage provides the basis for organization at the next succeeding stage. This does not mean, however, that all processes persist throughout life, nor does it mean that behaviors must remain stable across stages. To the contrary, development is essentially a dynamic process that promotes reorganization across time. More generally, development is probabilistic, not fixed by genes or by early experience (Gottlieb, 1983; Kuo, 1967).

---

[4]Baldwin (1897, 1915) was especially concerned that the methods available to psychology were not up to the task of correcting theory building.

4. Multiple levels of analysis are required because there are interlocked systems in the control of behavior. The investigator must be prepared to operate at different levels to achieve an adequate description of the phenomena. Kuo (1967) wrote that:

> the relationship between the behaving organism and its environment is an extremely complex and variable dynamic process. It goes deeper and beyond the molar level . . . *behavior* is far more than visible muscular movements. Besides such movement the morphological aspect, the physiological (biophysical; and biochemical) changes, the developmental history of the animal, and the ever-changing environmental context are interwoven events which are essential and integral parts of behavior. In our study of behavior, all such events must be investigated in a coordinated way. In other words, *the study of behavior is a synthetic science.* (p. 25)

5. Generalizing across species and across developmental stages requires comparisons on the basis of behavioral organization, not simple, salient characteristics (Jensen, 1967; Schneirla, 1966)[5]. To illustrate, Schneirla (1966) distinguished between seemingly identical acts on the basis of the social organization of the species. Specifically, biosocial organization (with a strong dependence on hormonal controls, such as insects) was sharply distinguished from psychosocial organization (with a strong dependence on psychological processes and learning, such as mammals).
6. Behavioral adaptation is a leading edge of biological adaptation, and the organism is continuously active and adaptive throughout the course of development, from birth to maturity and death. Developing organisms continuously assimilate features of the social and physical environment to themselves, and, in turn, accommodate their behavioral systems to those of other organisms.

A comparison of the six principles of the holistic orientation with the four common assumptions in cognitive developmental theory suggests that there is a high level of theoretical consensus. Is nothing fresh in the psychobiological perspective? Yes and no. Both orientations claim biology for inspiration and view issues of behavior and cognition in developmental, synthetic, and constructive terms. However, the model of developmental psychobiology differs from that of cognitive development on a basic dimension; namely, whether biology will be employed as metaphor or as science.

Although the two orientations use many identical terms—such as "progressive organization" and "hierarchical development"—they have different refer-

[5]Donald D. Jensen deserves credit for having provided a general framework for establishing the validity of cross-specific comparisons. According to Jensen, comparisons across species should be made on the basis of the biophysical, social, and developmental organization of which the activity is a part, not a single characteristic. The distinction is between *polythetic generalization* and *monothetic generalization*.

ents and bases for operational definition. In Piaget, for example, progressive organization and hierarchical integration typically refer to the stuff of cognitive schemata. He is concerned with the software of the mind. In Schneirla, holistic and associated terms refer to the totality of the individual processes—biochemical, behavioral, morphological, perceptual, and cognitive events. His developmental model combines biological form and behavioral functions; the hardware and the software. This is not an instance where different words are used to refer to the same thing. It is a bigger problem: The same words have been used to refer to different developmental processes. It is the scope and inclusiveness of the holistic view that differentiates Schneirla's holistic developmental theory from cognitive-developmental theory.

In that spirit, I offer some thoughts on how a holistic view on the development of self-knowledge can be usefully integrated with the genetic epistemology of Piaget, Baldwin, and Kohlberg. Then I return to the linkages between a developmental perspective on self-knowledge and evolutionary epistemology.

Self-concepts and Social Consensus: Two Masters

According to Quine (1981), there coexists a "drive for evidence and the drive for system" in any good theory (p. 31). Rather the same observation can be made about the "theories" that persons develop for themselves. Self-constructions, and cognitions of the world outside the self, must serve two masters. One is the need to achieve "system" or internal synchrony; the other is to achieve external validity or correspondence with "evidence." Consider first the internal coherence that permits the rest of the biological organization to function in an efficient, adaptive fashion. In the absence of synchrony within one's thoughts of oneself and of the world beyond, the organismic system may itself be challenged beyond appropriate limits. Accordingly, the biological system has a strong bias toward internal consistency and equilibrium among its components. Among other things, this bias generates pressure for internal coherence and harmony among thoughts and biology. In Freudian terminology, the integrity of the ego is protected by defense mechanisms.

The second need in one's theory of oneself is to achieve empirical confirmation.[6] Implicit in recent views of social-cognitive development is the assumption that mature and well-adjusted persons should have a convergent sense of themselves and other persons. According to this view, evaluations by persons from without should be consistent with the individual's own assessments from within. The correspondence assumption is in line with a commonsense view of the relations between social reality and the constructions that it imposes on persons. This view is also consistent with the neobehavioristic explanation for the emergence of self-ideas, in that primacy is assigned to the environmental control of self-concepts (Bandura, 1982; Cooley, 1922; Mead, 1934).

[6]The following two sections borrow heavily on passages from Cairns and Cairns (1988).

But the empirical findings available indicate that self–other correspondences are not inevitable in the course of development (Cairns & Cairns, 1988). The evidence fails to show increasing levels of agreement between the self and the social consensus about the condition of the self. The more global and private the dimension of evaluation, the less the level of correspondence (Kenrick & Stringfield, 1980). In this regard, the social realism embodied in the correspondence assumption is only one of the epistemological alternatives available. Another solution to this version of the body–mind problem would start with the analysis of functions during development rather than with its structure at maturity. In a different context, we have proposed that self-cognitions do not always have to be veridical in order to the functional (Cairns & Cairns, 1981, 1988; see also Epstein, 1973; Greenwald, 1980; Kelley, 1973; Wallwork, 1982). Even if one is sick and anxious and poor, there should be reason to get up in the morning. The essential idea is that one's view of one's self may serve different functions from those served by one's view of others.

Psychobiological Integration in Self Concept Development

Within our perspective on developmental epigenesis, internal synchrony and adaptive self-concepts do not automatically presuppose high levels of accuracy with external information sources (Cairns & Cairns, 1988). But should there not be increasingly higher levels of self–other correspondence as the child grows older, more experienced, and more perceptive? Not necessarily. Although cognitive development clearly implies greater complexity, it does not imply better social adjustment, higher degrees of ethical conduct, or increased levels of self–other agreement.

This is where the Schneirla agenda on development is of great importance. Within a holistic model, developmental change is not limited to cognitive modifications. At puberty, there are age-related effects on biophysical status—including physical and hormonal growth—which are more dramatic for social adaptation than modifications in cognitive competence. Indeed, adolescence refers primarily to changes in physical and sexual capabilities, not to changes in cognition. Similarly, the major transitions in mid-life are associated with biophysical changes in function and social roles, not sharp changes in cognition. In the context of the organismic perspective, it would be reasonable to expect cognitive conceptions of the self and others to be integrated with basic biosocial and sexual age-related changes at each developmental stage.

Four types of ubiquitous change in childhood and adolescence may be identified: the ability to think abstractly, the ability to communicate effectively and deceptively, sexual and social dimorphism, and modifications in social organization and social roles. Each of these developmental shifts has implications for one's concept of self, and its relation to the concepts that others share about him or her.

Abstractness and rationalization. There is an age-related increase in the cognitive ability of individuals to extend beyond the self, to perceive and to empathize with the motives and actions of other persons (Harter, 1983; Lewis & Brooks-Gunn, 1979). More generally, there is an increase in the capability of adolescents and adults to employ abstract concepts in describing the functioning of other persons, their own feelings and concerns, and the social ecology. A by-product of this cognitive advance is not only an enhanced complexity in understanding, but an age-related advance in the ability to invent both compelling explanations for the acts of others and convincing self-deceptions. Rationalization works by reconstructing the present and reorganizing the past; cognitive advances provide distance between the here-and-now concrete features of existence and the person's conceptions of him or herself and the social world. Whatever advances for veridicality that are won by a more precise discrimination of the external world may be lost by the creation of abstractions in which the details are submerged or redefined.

Biophysical Changes and Motivation. Biophysical, emotional, and social growth occurs from infancy to adulthood. Certain changes are universal to members of the culture of which the person is a part, but variations in the rate and content of the changes are distinctive to the individual. They lead to inevitable modifications in social needs and social status. For instance, there are age-related changes in the actual structure of the social systems in which children, adolescents, and adults participate, and in the social roles they fill. Children's peer clusters are typically unisexual; by mid-adolescence they are heterosexual. There are changes as well in biophysical status and motives associated with these changes. One of the more obvious is the onset of sexual and romantic concerns in adolescence, and the correlated changes in the heterosexual social structure and personal relationships. Similarly, new levels of physical maturation lead to changes in social status and responsibility at several critical points in life. These stages recur from early childhood and adolescence to maturity, middle age, and senescence.

Communication Networks and the Pollyanna Bias. There are age-related increases in the extent to which one may learn what other persons think about oneself directly from others. Communication about other persons is enhanced in large measure by changes in one's cognitive capabilities, on the one hand, and freedom of movement across social groups and social relationships, on the other. To the extent that one's social cognitions are shaped by communication with others in a social network, attention must be given to one's associates and how these are related to one's developmental and social circumstances. Furthermore, "Pollyanna biases" operate at virtually all developmental stages in communications about the self. Social protocol (i.e., etiquette) ensures that most interchanges are not "rude" or provocative, hence it is not polite to mention information that may be disturbing to the other person, however true it may be.

The same sanctions do not apply to information given to others about the person when they are not aware (i.e., gossip). Skills in social concealment and skills in social detection increase with age, at seemingly the same rate.

Social Ecology and Social Roles. Conjoined to changes in the social network are certain inevitable changes in one's social role and the expectations associated with that role (Bronfenbrenner, 1979). At 16 years of age, kids can buy cars, get jobs, obtain guns, get pregnant, and drop out of school. In the CLS longitudinal study (Cairns & Cairns, 1990), 75% of the teenage subjects have access to guns, the majority own cars and work while going to school to pay for them, and a few are already in jail or have died because of violence and accidents. Small wonder that their views of themselves and other persons have undergone changes since mid-childhood. Much of the content of these roles at the various stages in living is defined by the social culture that, in turn, becomes incorporated into the individual's values and perspectives.

The Correspondence Assumption and Self-Knowledge

Of these two functions of self-cognitions, I would propose that the primary one is the preservation of self-integration and direction. It is primary both in ontogeny and personal adaptation. The "holistic" assumption is embedded within the nuclear theories of cognitive development (Baldwin, 1897, Piaget, 1952) and developmental psychobiology (Kuo, 1967; Schneirla, 1966). Internal synchrony—whether in physiology or cognition—is not something that must be socialized or learned. It is a starting point for ontogenetic accommodation and evolutionary adaptation. According to modern theories of development, internal synchrony is present at the outset of development, though it may be challenged by changes from within and from without. Moreover, despite age-related modifications in role and capabilities, the bias toward within-system integrity and balance should remain dominant.

There is one central point of unclarity in "holistic" models of development. Precisely what stuff should be integrated in development? On the one hand, Baldwin, Piaget, and others who have followed their lead have concerned themselves primarily with the organization of mental phenomena, including the organization of self-ideas. On the other hand, Kuo, Schneirla, and the developmental psychobiologists have dealt primarily with the integration of physiology and social behavior in development. To the extent that mental and biological domains are kept separate, this version of the body–mind problem may be skirted by parallelism or epiphenomenalism. But if the domains are permitted to interact, inevitable problems of causation and primacy arise.

Furthermore, the fact that some self-conceptions resist conformity to the social consensus suggests that there must be other, internal criteria for determining their establishment beyond social consensus; that is, the standards extend beyond those of the "looking glass" of Cooley (1922). Cognitions that the person holds about the self and others may be sufficient in themselves, to the extent that these cognitions promote internal integration and direction.

The natural history of human beings suggests that there should be substantial veridicality among persons in their interpersonal evaluations of other persons. The ability to distinguish between one's friends and enemies—persons one should trust and persons one should fear—is a primitive function of social cognition in everyday life. Social attributions take advantage of an evolutionarily stable strategy that appears relatively early in social-cognitive development. One's personal survival depends in large measure on accurate recognition of persons and things that are likely to produce threat, harm, or injury.

Whatever services concepts of the self and of others perform, they should be reasonably helpful in correctly diagnosing situations and persons to avoid, and situations and persons one can trust. In addition, commonality in social perceptions on these basic matters is a prerequisite for adequate communication and continuing affiliations with other persons. To the extent there is a basic confirmation and balance in these perceptions, one might further share in the information network. Conversely, to the extent there are discrepancies on these matters, there may be corresponding gaps in communication skills and discrepancies in information transmission. Accordingly, children and adolescents normally develop the ability to observe, communicate about, and weigh information about other persons so as to predict behaviors directed toward themselves.

One reason that a developmental increase in cognitive functioning does not always lead to greater veridicality across the self–other-reality domains is that the same abstract capabilities that increase skills in the perception of others may be employed to counterbalance those perceptions, through the construction of rationalizations and self-enhancing fictions. Given the balance between increments in abstractness, on the one hand, and increases in rationalization, on the other, across-the-board increments in self–other agreement seem unlikely. Hence it appears that these discrepancies will not be erased with age or advances in cognitive skills. Some of the reasons for self–other discrepancies in self-evaluations in adulthood and adolescence are the same as those in earlier developmental stages: Namely, my thoughts of myself serve different purposes than my thoughts of others. This principle is not age relative, and it should hold across the developmental course. What changes over time is the ability of the self to share in aspects of the broader communication network, and in one's ability to empathize, generalize, and rationalize.

Self-Knowledge and Evolutionary Epistemology

The origins and functions of consciousness were central to comparative psychology in its founding by George Jean Romanes (Romanes, 1884). From time to time, they have re-emerged as central issues in the discipline (Crook, 1980, Jaynes, 1976, Kummer, 1971; see Gottlieb, 1979, for an excellent historical summary). The science early encountered basic difficulties in an effort to integrate the stuff of the mind with empirical observations of behavior. Resolutions to these difficulties range from the canon of Lloyd Morgan and the behavioristic manifesto of John B. Watson to the selective thinking of James Mark Baldwin and the cognitive maps of E. C. Tolman.

One of the boldest suggestions in the recent literature is that consciousness developed in human beings between 3,000 and 2,500 years ago (Jaynes, 1976). It is also one of the most thoughtful and provocative treatments of the issues. Whether it is 3,000 or 30,000 years, it does seem to be the case that self-consciousness has emerged very rapidly—in almost a millisecond of evolutionary time. Whatever may be the merits of such proposals on evolutionary causation, it should be noted that microevolutionary changes in the nervous system and its organization can be introduced rapidly, and with precision. Jaynes' idea on the recent evolution of cognitive functions may not be as farfetched as it originally seemed. Behavior is not structure, and modest shifts in the nervous system can produce multiple changes in adaptation. On this count, we have elsewhere demonstrated that organization in the nervous system and behavior is more susceptible to rapid microevolutionary change than has been generally believed (Cairns, 1986; Cairns, MacCombie, & Hood, 1983; Gariépy, Hood, & Cairns, 1988). And this behavioral change in two to four generations seems to have been made possible by genetic modifications of neurobiological metabolism (Lewis, Gariépy, Southerland, Mailman, & Cairns, 1988). Our proposal that behavior and cognition is the leading edge of evolutionary change seems entirely consistent with Campbell's proposal on "downward causation" (Campbell, 1974, this volume).

But why self-consciousness? Organisms other than humans seem to have gotten along quite well without it or survived on rudimentary and limited forms of it. Why should we as a species become burdened with another hazard to internal organization, given the debilitating consequences of mental dysfunction and failure (e.g., schizophrenia, paranoia, depression, anxiety, and extreme self-doubts)? It may be the case that the consciousness itself was a by-product of other adaptive achievements that demanded in human beings the ability to practice deceit and deception, and the ability to diagnose and respond to these actions. On this score, Kummer (1971) argues that cognitive skills to understand and control social processes came first in evolution, then "nonsocial" functions of intelligent tool use and symbolic planning followed. In the process of solving the problems of social organization (e.g., determining who were ones' friends and enemies beyond superficial appearance), new cognitive skills were evolved. Among these skills were those involved in actions in the external world, and in the ability to think about oneself, or to introspect. With introspection, there was also the necessity to formulate concepts of "theories" about one's own character and motivation. This loss of innocence in the evolutionary history of the species is duplicated in the ontogetic history of the individual.

TOWARD A REINTERPRETATION OF SELF-ESTEEM

A developmental and evolutionary perspective on the self has certain counter-intuitive implications for understanding self-esteem and its functions. It points toward a revision of the proposal that children and adolescents who are deviant in their social behavior are also deficient in social cognition in general and self-esteem in particular (Lewis & Brooks-Gunn, 1979; Harter, 1983). On the social-deficit view, children who mis-attribute motives and actions to others necessarily create difficulties for themselves which result in peer rejection and self-devaluation (e.g., Dodge, 1986). Proposals of social-cognitive deficit have since been applied to a wide range of adolescent problems, from explanations of conduct disorders and dropping out of school to learning disabilities and teenage pregnancy (e.g., Simmons & Blyth, 1987).

Although the social-cognitive deficit proposal has been persuasive, some issues remain to be clarified in both the model and the data that have been offered in its support (Cairns & Cairns, 1988). One problem is methodological. The evidence cited for the low self-esteem of children who fail in school and society has relied on information supplied by the individuals themselves. Typically they are asked to judge themselves with respect to such characteristics as their performance in school, their likelihood of getting into fights and arguments, and the level of their achievement on a variety of activities (sports, school, friendships). These self-evaluations are then compared to the self-evaluations of other children who have not shown a history of failure in multiple domains of living. The data show that children who are failures in life also tend to see themselves as being below the norm on several dimensions of adjustment.

So far, so good, except that self-report data may not be sufficient in themselves to conclude that children's self-esteem is inconsistent with their actual competence. Indeed, even low self-evaluations may be higher than the child deserves, relative to his or her observed behaviors. To determine whether children inflate or deflate their self-judgments relative to the judgments of other persons, beyond-the-self evaluations of the children are required. To the extent that the self-assigned scores match, exceed, or undervalue the evaluations assigned to the child by other persons, then self-esteem would be accurate, high, or low.

When self-other comparative assessments have been made, the outcomes indicate that seriously deviant children do not devalue themselves, nor do they differ from nondeviant children in the tendency to distort their self-descriptions. Our longitudinal studies of self-concept illustrate the point (Cairns & Cairns, 1988). We contrast the self-report measures of esteem with reports by other persons in three same-age groups: (a) highly at-risk boys and girls, (b) control children who were individually matched to the at-risk group on multiple dimensions, and (c) a general normal population. The work was replicated in two cohorts: 4th graders (N = 220) and 7th graders (N = 475). In accord with previous findings, both boys and girls who were at extreme risk for problem behavior tended to judge themselves as somewhat below the norm on multiple characteristics. Had we considered only these data, we would have concluded that at-risk children have low self-esteem as well as difficulties in getting along with other children and teachers.

But at-risk children typically judge themselves as being better adjusted than do other persons (Cairns & Cairns, 1981, 1984). In this investigation, the self-reports of both male and female at-risk subjects provided an inflated version of their competence and skills. They saw themselves in a positive light compared to how teachers and peers evaluated them, their behaviors, and their school performances (Cairns & Cairns, 1990). They judged themselves to be less aggressive, more popular, and more friendly than others judged them. This outcome is consistent with a function of self-ideas to buffer the private self from discordant public evaluations, even when the child is unloved and unloving. This buffer might be even more important for rejected children than for likable ones.

One other outcome deserves mention. The above self-other comparisons used algebraic differences in order to preserve the direction of the effect, whether inflation or devaluation. The assessment of self-other distortion (i.e., disagreement between self-evaluations and other-evaluations of the self, regardless of direction of the difference) is another matter. It requires the use of absolute self-other difference scores. Accordingly, the magnitude of the self-other distortion is obtained by summing the absolute self-other difference scores across subjects in each group. When absolute self-other difference scores are computed, virtually identical levels of distortion are obtained across samples. That is, at-risk, deviant youth obtained average absolute self-other difference scores that were similar to those of normal and control subjects (Cairns & Cairns, 1990). At-risk subjects tended to inflate their self-evaluations relative to others, while the control and general sample had both negative and positive self-other discrepancies.

To sum up, (a) at-risk children and adolescents provided somewhat lower levels of self-evaluation than nondeviant subjects, but (b) at-risk children had inflated views of their own capabilities, behaviors, and skills relative to the evaluations of others and (c) at-risk, control, and normal groups showed similar absolute levels of self-other distortion. These outcomes help explain the failure of interventions that attempt to reduce antisocial behavior by raising the children's self-esteem. The problem may be that deviant children already think too highly of their skills and behavior.

Why has it been so easy for the society and science to accept the dual propositions (a) that problem children have low self-esteem and (b) that self-devaluation and social-cognitive deficits lie at the root of their behavior problems? A possible answer may be found in the propensity to attribute to antisocial children themselves those beliefs and attitudes that others have of them. On this score, it appears that deviant children's self-esteem is not as low as the esteem in which they are held by others. Hence the social consensus and beliefs about the unworthiness of antisocial children may become falsely attributed to the problem children themselves. Once projected, these beliefs are then employed to explain their unacceptable behaviors.

Such social projection and tautological reasoning have not been limited to theories about deviant, antisocial children. They appear in the naive expectation that old, destitute, black women who are widowed in the inner city should have extremely low levels of life satisfaction (Mills-Byrd & Cairns, 1987). Exactly the opposite is the case. Despite their abysmal circumstances, such disadvantaged women have self-concepts and views of their lives that are positive and functional for their everyday adaptation. A similar principle seems to operate in the lives of children and adolescents, even among those who are rejected and devalued by others.

CONCLUDING REMARKS

These inquiries into the developmental roots of self-knowledge have been guided by the comprehensive view of the organism provided by the holistic theory of developmental psychobiology. But there are a couple of twists in the model that should be noted. Our inquiries concern humans, not animals; and they deal with cognition and consciousness, not biology and behavior. I suspect that it might be useful if models of evolutionary epistemology were also to give greater attention to these issues of human cognitive development. Individual knowledge is influenced by but not reducible to the nature of nerve endings and/or the neurobiological apparatus that controls information processing. Knowledge is itself relative, created by the self in order to serve functions for the developing, adapting organism. Changes in the nature of the organism, and changes in the nature of its knowledge and its theory of reality, are inevitable over time. If evolutionary epistemology is applied to the human condition, attention must be given to these dynamic changes over the lifetime of the person. Naturalistic and evolutionary considerations can help clarify the nature and limits of knowledge; developmental considerations can assist in understanding how knowledge is constructed and changed by each person. A systematic epistemology should be informed by both.

ACKNOWLEDGMENT

It is a signal honor to contribute to a symposium to celebrate T. E. Schneirla's contributions to the science and the society. This research is a product of the CLS longitudinal project (Robert B. Cairns and Beverley D. Cairns, Co-Directors). The Spencer Foundation and NICHD (R01 23302) provided support for this work.

REFERENCES

Baer, K. E. von. (1828). *Über Entwickelungsgeschichte der Thiere.* Königsberg: Bornträger.

Baldwin, J. M. (1897). *Social and ethical interpretations in mental development: A study in social psychology.* New York: Macmillan.

Baldwin, J. M. (1915). *Thought and things* (Vol. 4). London: Swan & Sonnenschein.

Bandura, A. (1982). Self-efficacy mechanism in human agency. *American Psychologist, 37,* 122-147.

Bronfenbrenner, U. (1979). *The ecology of human development: Experiments by natural design.* Harvard University Press.

Cairns, R. B. (1979). *Social development: The origins and plasticity of social interchanges.* San Francisco: Freeman.

Cairns, R. B. (1986). An evolutionary and developmental perspective on aggressive patterns. In C. Zahn-Waxler, M. Cummings, & S. Iannotti (Eds.), *Altruism and aggression: Social and biological origins* (pp. 58-87). New York: Cambridge University Press.

Cairns, R. B. & Cairns, B. D. (1981). Self-reflections: An essay and commentary on "Social cognition and the acquisition of self." *Developmental Review, 1,* 171-180.

Cairns, R. B., & Cairns, B. D. (1984). Predicting aggressive patterns in girls and boys: a developmental study. *Aggressive Behavior, 10,* 227-242.

Cairns, R. B., & Cairns, B. D. (1988). The sociogenesis of self-concepts. In N. Bolger, A. Caspi, G. Downey, & M. Moorehouse (Eds.), *Persons in social context:* Developmental processes (pp. 181-202). New York: Cambridge University Press.

Cairns, R. B., & Cairns, B. D. (1990). *Lifelines and risks: Adolescents in transition.* Unpublished manuscript.

Campbell, D. T. (1974). Evolutionary epistemology. In P. A. Schilpp (Ed.), *The philosophy of Karl Popper* (pp. 413–463, vol. 14, I & II). *The Library of living philosophers.* LaSalle, Il: Open Court.

Cooley, C. H. (1922). *Human nature and the social order* (rev. ed.). Glencoe, Il: Free Press.

Crook, J. H. (1980). *The evolution of human consciousness.* New York: Oxford University Press.

Dodge, K. A. (1986). Social information-processing variables in the development of aggression and altruism in children. In C. Zahn-Waxler, E. M. Cummings, & R. Iannotti (Eds.), *Altruism and aggression: Biological and social origins* (pp. 280-302). Cambridge: Cambridge University Press.

Epstein, S. (1973). The self-concept revisited: Or a theory of a theory. *American Psychologist, 28,* 404-416.

Flavell, J. H. (1977). *Cognitive development.* Englewood Cliffs, NJ: Prentice-Hall.

Gariépy, J-L., Hood, K. E., & Cairns, R. B. (1988). A developmental-genetic analysis of aggressive behavior in mice: III. Behavioral mediation by heightened reactivity or increased immobility? *Journal of Comparative Psychology.*

Gottlieb, G. (1979). Comparative psychology and ethology. In E. Hearst (Ed.), *The first century of experimental psychology* (pp. 147-176). Hillsdale, NJ: Lawrence Erlbaum Associates.

Gottlieb, G. (1983). The psychobiology approach to developmental issues. In P. H. Mussen (Ed.), *Handbook of child psychology* (Vol. 2, 4th ed.) (pp.1-26). New York: Wiley.

Greenwald, A. G. (1980). The totalitarian ego: Fabrication and revision of personal history. *American Psychologist, 35,* 603-618.

Harter, S. (1983). Developmental perspectives on the self-system. In P. H. Mussen (Gen. Ed.) and M. Hetherington (Ed.), *Handbook of child psychology* (pp. 275-386, Vol. 4, 4th ed). New York: Wiley.

Jaynes, J. (1976). *The origin of consciousness in the breakdown of the bicameral mind.* Boston: Houghton-Mifflin.

Jensen, D. D. (1967). Polythetic operationism and the phylogeny of learning. In W. C. Corning & S. C. Ratner (Eds.), *Chemistry of learning: Invertebrate research.* New York: Plenum Press.

Kelley, H. (1973). The processes of causal attribution. *American Psychologist, 28,* 107-128.

Kenrick, D. T., & Stringfield, D. O. (1980). Personality traits and the eye of the beholder: Crossing some traditional philosophical boundaries in the search for consistency in all the people. *Psychological Review, 87,* 88-104.

Kohlberg, L. (1969). Stage and sequence: The cognitive-developmental approach to socialization. In D. A. Goslin (Ed.), *Handbook of socialization theory and research.* Chicago: Rand McNally.

Kummer, H. (1971). *Primate societies: Group techniques of ecological adaptation.* Chicago: Aldine Press.

Kuo, Z.-Y. (1967). *The dynamics of behavioral development.* New York: Random House.

Lehrman, D. S. (1970). Some semantic and conceptual issues in the nature–nurture problem. In L. R. Aronson, E. Tobach, D. S. Lehman, & J. S. Rosenblatt (Eds.), *Development and evolution of behavior: Essays in memory of T. C. Schneirla* (pp. 17-52). San Francisco: Freeman.

Lewis, M., & Brooks-Gunn, J. (1979). *Social cognition and the acquisition of self.* New York: Plenum Press.

Lewis, M. H., Gariépy, J-L., Southerland, S., Mailman, R. B., & Cairns, R. B. (1988). Alterations in central dopamine pathways induced by selective breeding for aggression and social experience. *Society for Neuroscience* (Abstract)

Lorenz, K. (1962). Kant's doctrine of the *a priori* in the light of contemporary biology. *General Systems* (Vol. VII). Ann Arbor, MI: Society for General Systems Research.

Mead, G. H. (1934). *Mind, self, and society.* Chicago: University of Chicago Press.

Mills-Byrd, L., & Cairns, R. B. (1987). *Life satisfaction in elderly, poor, widowed, Black women.* Unpublished manuscript.

Piaget, J. (1926). *The language and thought of the child.* New York: Harcourt Brace. (originally published, 1923)

Piaget, J. (1952). *The origins of intelligence in children.* New York: International Universities Press. (Originally published, 1936)

Piel, G. (1970). The comparative psychology of T. C. Schneirla. In L. R. Aronson, E. Tobach, D. S. Lehrman, & J. S. Rosenblatt (Eds), *Development and evolution of behavior: Essays in memory of T. C. Schneirla* (pp. 1-13). San Francisco: Freeman.

Quine, W. von O. (1981). *Theories and things.* Cambridge, MA: Belknap Press.

Romanes, G. J. (1884). *Mental evolution in animals.* New York: Appleton.

Schneirla, T. C. (1966). Behavioral development and comparative psychology. *Quarterly Review of Biology, 41,* 283-302.

Simmons, R. C., & Blyth, D. A. (1987). *Moving into adolescence: The impact of pubertal change and school context.* New York: Aldine.

Wallwork, E. (1982). Religious development. In J. M. Broughton & D. J. Freeman-Moir (Eds.), *The cognitive developmental psychology of James Mark Baldwin: Current theory and research in genetic epistemology.* Norwood, NJ: Ablex.

Werner, H. (1940). *Comparative psychology of mental development.* New York: Harper & Brothers.

6 Paradigms Lost and Regained: Changing Beliefs, Values, and Practices in Neuropsychology

Stephan L. Chorover
Massachusetts Institute of Technology

OVERVIEW

The ambiguity of my subtitle is intentional. Its purpose is to epitomize the threefold aim of what follows, namely (1) to describe the epistemological, axiological, and methodological characteristics of the composite conceptual and material framework (or "paradigm"), within which most of what currently passes for the scientific study of mental and behavioral activities in my own field of academic and professional specialization gets done; (2) to locate the origins of this prevailing paradigm and to trace the process by which its constituent beliefs, values, and practices have attained the hegemonic status that they currently enjoy; and (3) to identify some of its limitations and to explore the prospects for its transformation into (or replacement by) a scientific paradigm possessing epistemologically, axilogically, and methodologically more sustainable conceptual and material characteristics.

INTRODUCTION

In setting us the task of considering epistemological and methodological issues in the study of mind, the organizers of this volume have invited us to become involved in some risky business. I am not trying to be melodramatic. Let me explain. In what Kuhn (1962) termed its *normal* phase, the conduct of scientific inquiry in any field presupposes an uncritical (perhaps even un-self-conscious) acceptance of certain beliefs, values, and practices. Like the trusting patient who faithfully follows the family doctor's orders, the practitioner of "normal science'' is loyal to the epistemological, axiological, and methodological dictates of the prevailing paradigm and declines to challenge the propriety of its presuppositions. Moreover, to such a practitioner, any such challenge is likely to appear unwelcome, especially insofar as it threatens to undermine the conceptual and material foundations of his or her scientific work.

But, when a remedy—however well established—turns out to be incapable of effecting a cure or is found to be exerting some dangerous and untoward side-effects, at least some of those who are in a position to do so start searching for more efficacious and safer modes of therapeutic intervention. By the same token, there are moments in any field of scientific inquiry (and thus, at least potentially in the life of any scientist) when even the most time-honored paradigm appears to have become inadequate to its task. In Kuhn's (1962) terminology, these are moments of *crisis*—literally *turning points.* In those sectors of the affected community where they are perceived as such, they engender an effort to achieve an appropriate epistemological, axiological, and methodological reorientation (i.e., a ''paradigm shift'').

It is obvious, of course, that we approach questions about the alleged adequacy or inadequacy of our scientific paradigms the way we approach questions of every kind: from a vantage point that is conceptually and materially conditioned and constrained by our own particular personal and professional experience. It follows that differences of opinion are bound to exist within any given field of scientific specialization, at any given moment in its development, regarding the epistemological, axiological, and methodological adequacy or inadequacy of the beliefs, values, and practices dictated by the prevailing paradigm (Gergen, 1985).

That being accepted, it also follows that the scientist who experiences a crisis of confidence in the adequacy of the paradigm prevailing within his or her own field, and who thereupon undertakes an examination of its underlying presuppositions, runs the further risk of being regarded as a renegade (or worse) by colleagues whose allegiance to the prevailing paradigm remains secure.

It seems, therefore, that a scientist who sincerely aspires to retain good standing in his discipline, yet feels conscience bound to challenge the epistemological, axiological, and methodological foundations of its prevailing paradigm, is likely to experience a modicum of what social psychologists call *cognitive dissonance.* That is my present predicament. Thus, in the hope of persuading my colleagues that what follows is not intended as an act of disciplinary disloyalty, I want to offer something in the way of a justification. My excuse is as follows:

Even with the benefit of hindsight, I find it impossible to say exactly how or where or when my present personal and professional perspective began to develop. But, I want to make it clear that my initiation into the epistemological, axiological, and methodological dictates of the presently prevailing (a/k/a ''modern'') scientific paradigm occurred quite early, and that my allegiance to its dictates was well and strongly established long before I became a neuropsychologist. Therefore, let me start by saying that I can recall thinking of myself as a scientist of sorts since my days at the Bronx High School of Science. In any event, I had been thinking of myself in those terms for quite a while before entering college and encountering my first real academic and professional mentor, Danny Lehrman. He taught me as an undergraduate, gave me my first exposure to T. C. Scheirla's ideas, helped me to realize that I wanted to become a neuropsychologist, and headed me, after graduation, to New York University. There, during the late 1950s while pursuing my graduate studies in neuropsychology under the direction of Hans-Lukas Teuber, I participated in Professor Schneirla's justly celebrated comparative psychology course.

I more fully acknowledge my debt to Professors Lehrman and Schneirla in due course. But the point that I make here and now is that the influences of Professors Lehrman and Schneirla notwithstanding, I am a neuropsychologist by training. The field in which I was trained is, as its name implies, a hybrid discipline. Its main goal, as I have always viewed it, is to comprehend relations between the organization and development of the central nervous system—and especially the human brain—on the one hand, and the organization and development of mental life and behavior, on the other.

Further to the point, throughout the initial phases of my scientific career and until relatively recently, I was far less interested in taking a "comparative"—as distinguished from a "physiological"—approach to neuropsychological research. Indeed, precisely because I was already fully committed to the epistemological and methodological perspective of analytical atomism, I was drawn to neuropsychology, not only to its subject matter but also to the idea of approaching it (as I saw it being approached) in accordance with the mechanistic and reductionistic conception of science that I had been brought up with from a relatively early age. In other words, from the outset, my fascination with neuropsychology was complemented by my conviction that mechanistic reductionism provided the conceptual and material wherewithal to attain its main goal. Although I probably would have been much surprised to be told this at the time. I can now see quite clearly with the benefit of hindsight how completely and uncritically I had come to accept mechanistic reductionism as the ONE AND ONLY fully creditable epistemological and methodological approach to take for all scientifically significant intents and purposes!

The consequences of this acceptance followed in due course. Let me explain. It was an interest in comprehending human mental life and behavior that had drawn me into neuropsychology in the first place. To be more precise, it was my ultimate aspiration to understand the neurobiological basis of human learning and memory. But, it is one of the main tenets of analytical atomism, and mechanistic-reductionism, that the best (indeed the only scientifically respectable) way to understand things is to divide them up or break them down into their smaller, supposedly simpler (and hence ostensibly more conceptually and materially manageable), constituent elements or parts. In view of the obvious ethical considerations limiting the scope and nature of neuropsychological research with human subjects, I chose to focus most of my attention on animal studies. Accordingly, throughout my time as a graduate student, and for many years thereafter, my main line of research was aimed at simulating, under relatively restricted and isolated laboratory conditions and in relatively reduced (and ostensible "simple") systems, the recurrent interplay of plasticity and stability that attends various aspects of human learning and memory.

Inexorably (and not unproductively) my career in neuropsychology followed its own progressively more and more reductionistic course. From early research on learning and memory in brain-injured patients (Chorover & Cole, 1961), I turned to the investigation of conditioning and memory consolidation in laboratory animals (Chorover & Schiller, 1965). But that was just the beginning. Gradually, the focus of my work fell even more narrowly on the neurobehavioral study of sensitization and habituation in the olfactory and limbic systems of rodents (Chorover & Deluca, 1975). Eventually, it reached a point at which most of what was going on in my laboratory involved the use of tiny microelectrodes to study minuscule electrophysiological and chemical events in single neurons and small groups of neurons in microscopically tiny bits of brain tissue in vitro (Chorover, 1976a). At about that point, I began to feel that my own scientific work had gone just about as far as it was going to go in that particular direction.

Coincidentally, I found my interest in limbic brain mechanisms (as I then referred to them) was drawing me into what has come to be called *the psychosurgery debate* (Chorover, 1976b; Valenstein, 1980). This turn of events is worth discussing in some detail, for it directly engendered the present inquiry.

It was during a violence-ridden period of heightened international and domestic turmoil—with the Indo–China war raging abroad, and urban rioting, antiwar demonstrations, and civil rights struggles raging at home—that the idea of using brain surgery as a means of controlling the thoughts, feelings, and actions of putatively disturbed, disruptive, or dangerous persons or groups made its most recent reappearance in the United States (Valenstein, 1973, 1980; Chorover, 1973, 1974, 1976b, 1979, 1980a, 1980b). Thus, in the aftermath of the particularly stormy series of ghetto uprisings that swept the nation during the summer of 1967, for example, the prestigious *Journal of the American Medical Association* published a letter from three prominent physicians affiliated with a leading medical school and hospital (Mark, Sweet, & Ervin, 1967). In it, a previously undiagnosed form of "brain disease" was proposed as a possible cause of the violent and assaultive behavior of at least some of those who participated in the riots. As it developed, the authors of the letter had already started using a previously untried form of psychosurgery (bilateral stereotaxic amygdalotomy) to treat what they alleged to be otherwise intractable forms of socially disruptive behavior (Mark & Ervin, 1968, 1970).

From the time I first learned about it, there were things that made me uncomfortable with this approach. Somehow, the idea of reducing a sociocultural phenomenon like violence to the status of a personal psychological problem traceable, in turn, to some putatively "underlying" neurobiological derangement and then proposing to treat it by destroying the ostensibly offending limbic brain region seemed to me to make about as much sense as proposing the destruction of core brain regions involved in the regulation of ingestive behavior as a way of dealing with the worldwide problems of human malnutrition and starvation.

At first privately, and later publicly, in both professional and popular forums, I began to examine the epistemological, axiological, and methodological entailments of psychosurgical theory and practice. Over a period of many years, I contributed my share to the large number of quite extensive and thoroughgoing studies of psychosurgical theory and practice that have appeared.

But, in the entire course of the debate, there is at least one point that has never been made explicit, to the best of my knowledge, by any of those who have tried (as I have) to criticize psychosurgery from a scientific perspective. When challenged to justify psychosurgery, its proponents and practitioners are inclined to proclaim, with evident sincerity, that what they advocate and do is completely rational. The point is that they are correct.

Thus, from the rational vantage point of anyone who unswervingly accepts, without qualification, the epistemological, axiological, and methodological entailments of the prevailing mechanistic and reductionistic paradigm—which is to say, from the perspective that most of us were brought up to regard as the only truly scientific approach to take for all serious scientific purposes—all aspects of human psychological existence (i.e., all thinking, feeling, and acting) ARE presumably reducible, for all theoretically and practically relevant intents and purposes to "expressions of the functioning brain" (Sweet, 1970).

Moreover, once that paradigmatic orientation has been adopted (as it so often is and has been) as the sole "sound scientific basis" for defining and dealing with human problems, it does indeed become entirely rational to view social problems as completely reducible, first to the status of psychological disorders or derangements, and thence to "expressions" of a neurobiologically malfunctioning brain.

I regret that it has taken me so long to do so, but I now clearly recognize that psychosurgery IS as rational as its proponents have always claimed it is.

And that's the problem. Lest I be misunderstood, let me explain by recalling what Albert Einstein once said about the unleashing of atomic power. He said that it had "changed everything save our modes of thinking" and that was why "we thus drift toward unparalleled catastrophes." Further to the point, he observed that "By painful experience, we have learned that rational thinking does not suffice to solve the problems of our social life. . . . We need a new way of thinking if humanity is to survive." I agree. So much for my apology.

Where did we ever get the idea, in the first place, that the rational, mechanistic, reductionitic, analytical, atomistic approach is—or ought to be—sufficient when it comes to defining and dealing with the world and its contents—including ourselves, and what Einstein referred to as "problems of our social life"?

One way of beginning to answer this question is through an exploration of the essential epistemological, axiological, and methodological entailments of what I have been calling *the prevailing paradigm.* For it is the source from which most of our present (putatively "scientific") beliefs, values, and practices have come. It's influence prevails not only in my own field but also—with some notable exceptions—throughout science as a whole. It marked the century-long struggle of psychology to become a "real" science (Boring, 1950; Howard, 1985; Koch & Leary 1985; Krasner & Houts, 1984; Peele, 1981) and remains a powerful conceptual and material force in all areas of the broader sociocultural context of which the modern scientific enterprise itself is but a part (Mendelsohn & Weingert, 1977).

THE "MODERN" SCIENTIFIC PARADIGM

As everybody knows, the reorganization of western society during the sixteenth and seventeeeth centuries had as its epistemological, axiological, and methodological corollary a decisive paradigm shift (somewhat imprecisely termed "the" scientific revolution; Hanson, 1958; Kuhn, 1962). In attempting to characterize the "modern" scientific beliefs, values, and practices that emerged in this shift, it is helpful to begin with the "classical" paradigm it displaced.

Within it, humanity and nature were conceived of as counterparts in an organic and purposeful whole. Moreover, from ancient through medieval times, and most emphatically within the more animistic and vitalistic versions of this paradigm, the inhabited earth as a whole (the Greek OIKOUMENE) was frequently likened to the body of a living organism often personified as Gaia (or Ge), the original pagan mother goddess, whose etymological traces show up in such terms as geography, geometry, geology, etc. (Sahtouris, 1985). This early image of humanity and nature had axiological and practical implications also. As Merchant (1980) has pointed out, it "served as a cultural constraint restricting the actions of human beings. . . . As long as the earth was considered to be alive and sensitive, it could be considered a breach of human ethical behavior to carry out destructive acts against it" (p. 3).

By contrast, the "modern" scientific world view and value system entails no such constraints, is predicated on a radical ontological and epistemological detachment of "man" from "nature," and takes as its avowed goal the domination of the latter by the former (Liess, 1972). Further, its emergence coincided with a period of commercial expansion, global exploration, and increasing exploitation of natural resources.

In this connection, it is informative to recall some of the ways in which the "founding fathers" of the modern scientific paradigm described its avowedly masculine perspective vis-a-vis putatively feminine nature.

In Francis Bacon's influential view, for example, there were "laid up in the womb of nature many secrets of excellent use." He perceived in science, moreover, an enterprise in which "human knowledge and human power meet as one" ("Nam et ipsa scientia potestas est") with the avowed goal of getting nature to yield up her secrets. Toward that end, nature was to be probed and "hounded in her wanderings . . . forced out of her natural state and squeezed and molded . . . put in constraint . . . bound into service . . . (and made a) slave." (These and many additional quotations along similar lines may be found—with pertinent commentary regarding their prototypically macho blend of sexual and violent imagery—in Easlea, 1981; Griffin, 1980; Liess, 1972; Merchant, 1980).

Rene Descartes was of similar mind. Indeed, what we encounter in Cartesian rationalism is an extreme masculinization of thought, with a proudly individualistic intellect cavalierly laying claim to worldly omnipotence. Thus, in the introduction to Part 4 of his seminal "Discourse," for example, Descartes (1955) proclaims his new "Method" to be a means by which we scientists might hope to "render ourselves masters and possessors of nature" (p. 6).

Cartesian rationalism provides the modern scientific paradigm with its principal intellectual foundations. Its overall epistemological, axiological, and methodological hallmark is a rigorously rational, deterministic, mechanistic, and reductionistic conception of the universe as a whole.

To Descartes, for example, the material world was a machine and nothing but a machine. In his view—which soon became virtually universal within the scientific community—all living organisms are likewise machines operating according to strictly deterministic mechanical laws that are (in principle, at least) completely amenable to discovery through the rigorous application of the epistemological and methodological tools and techniques of mechanistic reductionism and analytical atomism (Lasswitz, 1890).

Given that its avowed goal was the domination and control of nature and that its principal conceptual and material methods of procedure involved the persistent division and decomposition of the objects of its interest, it should come as no surprise that there is no proper place in the modern scientific paradigm for values that might restrict or constrain the abusive exploitation of nature for human ends. Indeed,what is perhaps most notable in this connection is not that the putatively purely rational power of modern science and technology has so often been developed and deployed for unscrupulously exploitative and frankly destructive purposes, but that proponents of the modern scientific world view so frequently and loudly proclaim the entire enterprise of scientific inquiry to be entirely devoid of any axiological dimensions whatsoever. Thus, the modern scientific world view is frequently advertised as uniquely capable of providing a "true picture" of the world and its contents because it is allegedly "value-free" in the sense of possessing such supposed epistemological and methodological virtues as "moral and ethical neutrality," and "objectivity."

How is this strange posture of purportedly virtuous detachment attained? Via the stipulation of an Archimedean point. Let me explain. The Sicilian scientist and wonder worker, Archimedes of Acragas (287–212 B.C.), is famous in part for his ultimately unsuccessful struggle to find a single conceptually and materially fixed, nonrelative, absolutely immovable position of rest or "vantage point" to serve as the completely secure point of departure for his own reasonings about the world and its contents. According to legend, he prayerfully entreated the gods for such a point because he believed that from it he could "move the world."

With due acknowledgment to Archimedes' quest, Descartes purported to ground his own philosophical system, in its entirety, on the fact of his own self-conscious existence. This he did, recall via his most famous assertion: I think, therefore I am (Je pense, donc je suis: "Cogito ergo sum"; Discourse, Part 4).

As Descartes' younger contemporary, Pascal, was quick to argue, it is anything but clear on what basis rationalty can justly predicate itself as Archimedean point. What is it (Pascal wondered) that entitles "raissonment" (cognition) to divorce itself from the whole of experience in ways that for example, deny "sentiment" (affect) a role in the establishment of a trustworthy starting point for all knowledge? The question (I would say) has yet to receive a scientifically satisfactory answer. But it is plain that once the predication, the divorcement, and the denial are accepted, it takes no more than a short mindstep more for the rational mind to reason its way to the point of imagining itself to be a completely immaterial, ethereal, "thinking thing" entirely separate and distinct from things and events with which it might cognitively deal.

Thus, the "cogito" set the stage for a number of additional assertions intended to set "man" ontologically and epistemologically apart from (and above) "nature" and "reason" epistemologically, and axiologically apart from (and above) "experience."

Moreover, once established, the supposedly mutually exclusive, dualistic categories of Cartesian rationalism began to ramify and redouble in ways that lent the seductive appeal to images of the disembodied mind reaching out to "grasp" (or "comprehend") and thereby take cognitive control of the vast and insensate clockwork universe "out there." Thus, for example, to Descartes' own divorce of mind from matter was soon added not only the separation of mind from nature but also Newton's fundamental distinction between "absolute space" and "absolute time" (Capek, 1961).

The significance of this detached, disembodied, Archimedean conception of mind for the development of the modern scientific world view cannot be overemphasized. It offers the only available support to an otherwise patently unsupportable "specular ideal" of scientific "detachment" according to which a human observer, supposedly having succeeded in detaching him/herself from things and events observed, is in a position to observe and describe them "objectively."

In concluding this brief recitation of the main conceptual foundations of the modern scientific paradigm, it is pertinent to reiterate that it remained the virtually unshakable foundation for the entire edifice of modern science until the first quarter of the present century. As such, the mechanistic/reductionistic perspective has exerted a profound influence throughout psychology and all its disparate sects during the past century and continues to do so today (see e.g., Koch & Leary, 1985). Indeed, not only within neuropsychology, but across the entire spectrum of psychological subfields, from laboratory-based experimental psychology to clinic-based applied psychology, and from empiricistic behaviorism to rationalistic cognitivism, its influence has been overwhelming. Not even psychoanalytic approaches have been exempt. Freud himself, for example, made a persistent and self-conscious effort to use Newtonian mechanics as the model for his conceptions and descriptions of intrapsychic phenomena. With evident pride—tinged, perhaps, with a touch of "physics envy"—he proclaimed himself and his followers to be "at bottom incorrigible mechanists and materialists" (quoted by Capra, 1982, p. 180).

Significantly, the period during which psychology as a whole, and neuropsychology in particular, began endeavoring to accommodate itself to the mechanistic and reductionistic world view and value system of Cartesian rationalism and Newtonian atomism coincides with the period during which physics—the "real science" that psychology has most often and most insistently attempted to emulate—was beginning to emerge from under the influence of the modern scientific paradigm and to enter, according to Toulmin (1982): "a phase of scientific thinking that differs from the science of the eighteenth and nineteenth centuries quite as significantly as the 'new philosophy' of the seventeenth century differed from the science of pre-Renaissance Europe" (p. 231).

To be more precise, during the first quarter of the twentieth century, physicists were forced by developments in their own discipline to undertake a fundamental and wide-ranging reassessment of the epistemological, axiological, and methodological adequacy of precisely those beliefs and values and practices that psychology was, meanwhile, undertaking to imitate. Indeed, the curious fact that this particular paradigm shift has been so largely ignored by psychology and its relatives among the human sciences becomes even more curious once it is realized that the reassessment was provoked by findings that forced the physicists to recognize scientific inquiry as an incurably human activity (Mendelsohn, 1977).

THE EMERGING PARADIGM

Although the actual beginnings of such things are ultimately impossible to determine, symmetry suggests that we begin this discussion, as we did the one in the preceding section, by referring to the epistemological and methodological status of the Archimedean point. In 1905, Albert Einstein remarked that "It is impossible by any experiment whatsoever to determine absolute rest." Indeed, this was his way of initiating a discussion which eventually led to the special theory of relativity. For present purposes, this statement is important for its role in initiating a period of scientific inquiry in which human observers can no longer pretend to occupy positions radically separate and distinct from things or events observed.

The strange doctrine according to which scientists are—or ever could be—epistemologically, axiologically, and methodologically "detached" observers and interpreters of the objects and events that they undertake to comprehend must be abandoned and replaced with a recognition that observation entails an unbreakable relationship between observers and things or events observed. Thus, as investigators, we interact with the systems we observe. Accordingly, influences deriving from our personal and professional points(s) of view and frame(s) of reference can no longer be ignored. Indeed, an account of the observer's relation to the system must now be included as part of any purportedly scientific description of the system.

In those contexts to which it is applicable, the advent of this relativistic and relational mode of conceptualization renders wholly untenable the Cartesian subject/object dichotomy by replacing the "detached observer" pursuing the "specular ideal" of "objective knowledge" with an irreducibly compound observing/observed system-in-relation (Wheeler, 1982).

But that is only the beginning. The essentially relational and interactive nature of knowledge and knowledge making under the aspect of the emerging paradigm is made even clearer by a recognition of the fact that contexts exist in which systems cannot be observed unobtrusively or without altering them. In such contexts, it becomes untenable to even speak of "observation." Rather, it is necessary to acknowledge each act of investigation as an irreducibly composite instance of participation/observation, in which the observer's intervention affects the system being observed in ways that preclude the possibility of ever being able to say anything about the latter as a "ding an sich."

No doubt the most dramatic and mathematically well-formulated demonstrations of this fact come from research at the theoretical and experimental limits of twentieth-century high-energy physics. There, in the intracosmic, subatomic realm of the extremely small, the energy required to observe a system has been shown to interact with the energy inherent within the system in ways that make it disconcertingly impossible to describe the system independently of the observation process. This finding, formulated most precisely in Werner Heisenberg's (1972) justly famous "principle of uncertainty" has the effect not only of further implicating the knower with the known, but also of setting a limit to what is knowable.

There is a story that a physicist friend of mine likes to tell that may serve to introduce yet another aspect of the emerging paradigm. The story concerns a late teacher of his who studied physics during the critical first decade or two of the twentieth century. One point of contention, among the many that marked the crisis in physics at the time, was how to conceptualize and define various fundamental phenomena.

With respect to one such phenomenon, the photon, the community of physicists was about equally divided between those who regarded it as a particle and those who regarded it as a wave, and the young physicist in the story was supposedly pursuing his studies concurrently under two teachers of contrary persuasions on this fundamental question. How did he deal with this situation? "I did the only reasonable thing," he is supposed to have said, "on Mondays, Wednesdays, and Fridays I worked with Professor X and he convinced me that it was a particle; on Tuesdays, Thursdays, and Saturdays, I worked with Professor Y, and he persuaded me that it was a wave; on Sundays, I went to church and prayed for divine guidance."

As we all know, it took almost 20 years to resolve this issue. Gradually, however, the community of physicists accepted as a basis for their work at the atomic level of organization a conceptual framework that accorded equal legitimacy to descriptions pertaining to BOTH the "wave-like" AND the "particle-like" properties of photons and other elementary phenomena. This they eventually did, moreover, in accordance with Niels Bohr's "principle of complementarity," which has been characterized as follows (Oppenheimer, 1950):

> The basic finding was that in the atomic world it is not possible to describe the atomic system under investigation, in abstraction from the apparatus used for the investigation, by a single, unique, objective model. Rather a variety of models, each corresponding to a possible experimental arrangement and all required for a complete description of possible physical experience stand in a complementary relation to one another, in that the actual realization of any one model excludes the realization of others, yet each is a necessary part of the complete description of experience in the atomic world.

In a moment we must concern ourselves with the question of whether the findings of the kind that led Bohr to enunciate the principle of complementarity (as well as the findings that led to the formulation of the principles of relativity and uncertainty) are limited to the atomic world or whether they also have their counterparts in the study of systems at other levels or organization.

But, before doing that, let me make clear my main reason for retelling the foregoing story. It has nothing to do with waves and particles as such. Rather it has to do with the question of how such a controversy could have lasted for almost a generation, with both sides steadfastly refusing to accept a settlement essentially consistent with both positions. Why, in other words, did the community of physicists, faced from the start (as they obviously were) with the possibility of resolving this particular dispute in the manner in which it was eventually resolved, find themselves compelled to continue to argue with one another, back and forth, in categorical "either/or" terms? What prevented them from seriously considering the possibility of there being a plurality of equipotentially adequate ways to describe an aspect of the world as fundamental as a photon? Why were they so unwilling to let go of the presupposition that there could only be "one true" description? No doubt there are many equipotentially plausible answers!

In my own view, it is impossible to exaggerate the influence of "Archimedean-like" Cartesian presuppositions, coupled with a commitment to dualistic habits of rational thought and an uncritical acceptance of categorical ("either/or") logic. The scientists involved were people who had been brought up, after all, as most of us still are, in a problem-posing and problem-solving climate where it is not customary to question the adequacy of rational thinking for all serious scientific intents and purposes.

Bohr's idea that it is necessary to include BOTH wave-like AND particle-like properties of the photon in the picture of the physicist's relation to the subatomic world complemented the work of Einstein and Heisenberg by calling into question what is perhaps the single most fundamental proposition of modern science: the proposition that there is (or could be) only a single correct way of conceptualizing and describing relations between observers and things or events observed. Taken as a whole, their contributions serve to give shape to an emerging scientific paradigm that among other things, restores a host of connections between and among the entities and events that its predecessor insisted on regarding as essentially separate and distinct. It recognizes, for example, the ultimate continuity of spacetime, the fundamental interrelatedness of matter and energy, and the comparably inescapable epistemological and methodological involvement of the knower with the known.

So the emerging scientific paradigm differs from its "modern" scientific predecessor in a number of epistemologically, axiologically, and methodologically significant respects. For example, rather than purporting to provide a single, absolutely complete, veracious, and objective "picture of nature," the emerging paradigm, through its principles of relativity, uncertainty, and complementarity, offers the somewhat more modest prospect of providing a number of less complete and more interactively qualified pictures "of our relation to nature" (Heisenberg, 1958, pp. 58, 81).

Now, there is little reason to doubt that this paradigm, with the decisive epistemological, axiological, and methodological reorientation that it entails, is applicable to the study of systems at the scale of the extremely large and exceedingly small (systems, that is to say, at the cosmological and atomic levels of organization). Is there any reason to believe that it is applicable, mutatis mutandis, to scientific inquiry in the realm of human affairs as well? This is the question to which we now turn. As might be expected, the answer arrived at is generally affirmative.

Let J. Robert Oppenheimer set the ground tone. In the following passage, written more than four decades ago, he was speaking with specific reference to the principle of complementarity:

> It is of course not yet fully clear how characteristically or how frequently we shall meet instances of quite close analogy to the complementarity of atomic physics in other fields, above all in the study of biological, psychological and cultural problems. Yet it is clear, as has repeatedly been stressed by Bohr himself, that the discovery of complementarity has provided us with a far wider and more sophisticated framework for the synthesis of varieties of scientific experience. It has refined and extended the pluralism natural to science and added new elements of subtlety to the idea of dialectic. Indeed, it seems to offer a far richer and more adequate general point of view for the comprehension of human experience than the misleadingly rigid and unitary philosophies that flowed so naturally from the experiences of Newtonian mechanics. (ibid)

REVISIONING NEUROPSYCHOLOGY

All of us are aware of the perils that attend unwarranted transphyletic extrapolation in the study of animal behavior (Maier & Schneirla, 1935). These seem to have their counterparts in this discussion of scientific paradigms. Thus, granting the existence of a crisis involving the prevailing paradigm in neuropsychology, and granting the fact that a paradigm shift has been occurring in physics, borrowing from the latter to resolve the epistemological, axiological, and methodological problems of the former is unwarranted, unless or until grounds for so doing are found within our own fields of scientific activity (Edelman, 1978). Happily, they are not hard to find.

For example, as soon as we look at psychology with the foregoing considerations in mind, we encounter a scientific community deeply divided on the evidently fundamental question of what its proper subject matter is (Chorover, 1985). The particular line of cleavage I presently have in mind is between the "mentalists" or "rationalists" (on the one hand) for whom psychology is (in William James' (1890/1950) phrase) "the science of mental life" including "feelings, desires, cognitions, reasonings, decisions, and the like" (Vol. 1, p. 1), and the "behaviorists" or "empiricists" (on the other hand) for whom psychology is "the study of behavior" (Watson, 1914; Skinner, 1938).

Given the form and content and ultimate resolution of the wave/particle dispute that divided our colleagues in physics for so long, does it not seem obvious that the principle of complementarity offers a scientifically sustainable way of reconciling these two ostensibly irreconcilable and mutually exclusive alternatives?

Indeed, there are more substantive contexts within psychology where the perspectival implications of Einsteinian relativity apply, where probing presents problems analogous to those that led Heisenberg to propound his principle of uncertainty, and where dialectical (both/and) modes of conceptualization comparable to those embodied in Bohr's principle of complementarity can help to reconcile issues long stalemated in sterile categorical (either/or) debate. One such debate is presently going on around the question of whether mental "primacy" should be accorded to cognition (Lazarus, 1984), or to affect (Zajonc, 1984). But perhaps a more immediately pertinent example is the longstanding and periodically recurrent contest between advocates of "biological" and "environmental" determinisms. In its archetypal form—that is to say, as the infamously venerable "nature versus nurture" debate, this old chestnut is among us, once again; this time in the guise of the so-called "sociobiology controversy."

It was from Professors Lehrman and Schneirla—and, more particularly through their critical discussions of biological and environmental determinist thought in comparative psychology (e.g., Lehrman, 1953; Schneirla, 1954)—that I first learned to recognize the extreme scientific inadequacy and potentially profound social implications of the tendency to conceptualize and describe psychological phenomena as if they were completely reducible to expressions of insensate forces deriving from EITHER innate (e.g., genetic) OR acquired (e.g., environmental/experimental) sources.

Indeed, Lehrman's compelling studies of reproductive behavior in the ring dove (e.g., 1962, 1964) have always stood, in my own mind, as a most eloquent demonstration of the need to take the recurrent interplay of internal (e.g., hormonal) and external (e.g., environmental) influences properly into account in attempting to understand the organization and development of behavior patterns on which individual and species survival depend. Further, I have long admired Schneirla's concept of "integrative levels" (1949, 1957) for its ability to clarify relations between the continuities and the discontinuties encountered in the comparative psychological study of mental and behavioral activity.

But it was only recently, after spending many years working within the mechanistic and reductionistic framework of contemporary neuropsychology, after participating in the psychosurgery debate and the sociobiology controversy (Chorover, 1979), and after undertaking the present line of inquiry into the epistemological, axiological, and methodological entailments of the prevailing paradigm, that I found myself beginning to understand, in a new light, what those two teachers of mine were up to. I am happy to have this opportunity to describe my impressions in this regard and thus to acknowledge their influence on my own thinking.

I don't know whether or not things going on concurrently in physics had anything to do with it, and I don't have any reason to think that Professor Schneirla was the first or only one to come to the realization that categorical (either/or) modes of conceptualization are simply inadequate to the task of making the organization and development of mental and behavioral activities scientifically comprehensible (several Gestalt psychologists, for example, appear to have been concerned with similar problems, at around the same time). But I do see Professor Schneirla, together with his co-workers and students, as second to none among comparative and physiological psychologists in the appropriateness of their response to this realization. Indeed, in Schneirla's concept of integrative levels, no less than in Bohr's principle of complementarity, I see the dialectical mode of conceptualization being applied to fundamental scientific problems of broad scope in a most compelling, contextually appropriate way.

In drawing attention to this aspect of their likeness, it is my broader intention to point at what appears to be the more general epistemological, axiological, and methodological significance of the dialectical mode of thinking that they exemplify. Indeed, what I am trying to suggest is not only that this mode of thinking, when properly employed, is capable of providing a scientifically sustainable means of transcending impasses created by the collision of otherwise irreconcilable categorical perspectives, but also that it deserves to be relied on for this purpose more than it has been in my own field. By way of illustration, let me mention briefly some of the epistemological, axiological, and methodological dictates for the emerging paradigm, in general, and the dialectical mode of conceptualization, in particular.

As noted previously, a principal aim of my current work is to devise a scientifically-sustainable alternative to the presently prevailing neuropsychological paradigm. Within the mechanistic and reductionistic framework of the latter, mental and behavioral events tend to be viewed as if they exist and arise in a neuropsychological vacuum. By contrast, within the broader ecosystemic framework I am trying to develop, the organization and development of human systems is seen in composite synchronic (contextual) and diachronic (evolutionary) perspective (Chorover & Chorover, 1982; Chorover, 1986, 1987). In its present form, the framework bears a resemblance to one being developed concurrently by Wapner (1983, 1987). As already described, it endeavors to encompass the study of relations among the three main Aspects and Levels of organization shown in Table 6.1.

Thus, for example, rather than regarding the mental activity and behavior of individuals (i.e., personal "thoughts," "feelings," and "actions") as merely or even mainly "expressions" of "underlying brain mechanisms" (and, hence, conceptually and materially reducible thereto), I presently prefer to think of them as specifically psychological manifestations of a threefold (cognitive/affective/expressive) mode of organization that is characteristic of human systems, in general. As such, I am interested not only in their specifically psychological form and content and interaction, but also in their relations to what I conceive of as their cognitive, affective, and expressive counterparts at "underlying" neurobiological and "surrounding" sociocultural counterparts (Chorover & Chorover, 1982). From this vantage point, neuropsychological and social psychological phenomena arise and exist in the "borderlands" where psychological phenome-

TABLE 6.1
Human Systems: Levels and Aspects of Organization

	Neurobiological (organism)	*Psychological (Individual)*	*Sociocultural (Community)*
LEVELS:	*Brains*	*Persons*	*Groups*
Generic		*Specific*	
cognitive	neocortex	thinking	world view(s)
affective	limbic system	feeling	value system(s)
expressive	core	acting	life style(s)
DOMAINS:	*—Neuropsychology—*	*—Social Psychology—*	

Legend: Three principal aspects of human systems (cognitive, affective, and expressive) as manifested at three successive levels of organization (neurobiological, psychological, and sociocultural). Note that aspects are conceived of as having counterpart representations at different levels, and that the absence of lines between adjacent "cells" of the table is intended to imply the existence of extensive interplay among/between/within the indicated levels and aspects of organization. Note, also, that the epistemological, axiological, and methodological entailments of scientific paradigms being discussed here can be viewed as cognitive, affective, and expressive aspects, respectively, of the specifically human system that is science.

na interact and overlap and combine with neurobiological ones (on the one hand) and sociocultural ones (on the other).

This is obviously not the place to offer additional justifications for my decision to take this approach, or to enter into a detailed discussion of the research directions in which it has been taking me. However, it is an appropriate place to illustrate, by way of a single pertinent example, how misleading the mechanistic and reductionistic mode of neuropsychologizing can be when applied to the problem of comprehending the organization and development of mental and behavioral events on which individual and species survival depend, and how the adoption of an alternative dialectical perspective helps to revision neuropsychology by taking the interplay among events at the neurobiological, psychological, and sociocultural levels of organization more properly into account. An example directly related to the aforementioned psychosurgery debate makes the point.

When first performed on monkeys maintained under conditions of relatively solitary confinement in laboratory cages, bilateral amygdalotomy was found to produce marked "tameness" as well as several other mental and behavioral signs and symptoms that had been described earlier in laboratory-maintained monkeys who had undergone bilateral temporal lobectomy an that, taken together, comprise the so-called "Klüver–Bucy syndrome" (Klüver & Bucy, 1939).

In the mid-1960s, it was this evidence—coupled with psychosurgical reasoning of the kind that I have already described—that was cited as providing ostensible scientific justification for the subsequent attempt to use bilateral stereotaxic amygdalotomy as a way of "taming" people who were allegedly wild and violent and unmanageable (Chorover, 1976b).

However, as early as the mid-1950s, it was already clear that tameness was not observed following bilateral amygdalotomy in monkeys who were maintained as members of a group, within a relatively spacious but nonetheless confining laboratory compound. Instead, brain lesions of the kind we are discussing produced striking and generally unpredictable bipolar changes in emotional reactivity and patterns of social interaction (Rosvold, Mirsky, & Pribram, 1954).

Further, when the same surgery was subsequently performed on monkeys who had previously been living under somewhat more normal conditions—among others in a relatively free-ranging island habitant—the "tameness" that they showed during a period of postoperative confinement in the laboratory was replaced by social ostracism, starvation and eventual death when they were returned to their former environment (Kling, 1972).

Thus, it is unwarranted to make any single, simple, straightforward, and otherwise unqualified "cause-and-effect" assertion about mental and behavioral aftereffects of bilateral amygdalogomy. Indeed, linear causality of this kind provides at very best a partial, incomplete, and potentially misleading way of thinking about neuropsychological relations. In fact, the results observed following bilateral amygdalotomy in the monkey turn out to be exceedingly complex and context dependent, in the sense that they are decisively conditioned and constrained by a number of factors including, most pertinently, the nature of the environing social situation in which the animals are maintained and studied. Nor is the context dependency illustrated by this example a rare or unusual occurrence. Many similar instances could readily be adduced from various areas of neuropsychological research.

What conclusions can we draw? What are the implications of all this for our understanding of neuropsychological relations in particular, and for the scientific study of human systems more generally?

In closing, let me offer some provisional answers from my own present personal and professional perspective, and in a form intended to echo, to some extent, Oppenheimer's remarks quoted previously. In the realm of neuropsychological research (as in the atomic world), it is not possible to describe the system under investigation, in abstraction, apart from the context of investigation. The latter includes, but is not necessarily limited to, influences attributable to the investigators point-of-view and frame-of-reference. Possible effects of the investigative process on the behavior of the system under study must not be ignored or overlooked. The existence or attainability of any single, unique, complete, objective description, interpretation, or model or neuropsychological relations should not be presumed. Rather, insofar as the actual realization of any one description, interpretation, or model excludes the realization of others; and each is a necessary part of any scientifically sustainable description, so a variety of descriptions, interpretations, and models, each corresponding to a particular experimental arrangement and all required (in lieu of any humanly attainable complete description of all possible neuropsychological relations) stand in complementary relation to one another in any scientifically sustainable description, interpretation, or model of neuropsychological relations.

As it becomes increasingly clear that we characteristically and frequently meet instances of quite close analogy to the complementarity, uncertainty, and relativity of atomic physics in other fields, and above all, in the neurobiological, psychological, and sociocultural study of human systems, we must be prepared to approach the study of such systems from within the conceptual and material framework of an epistemologically, axiologically, and methodologically sustainable paradigm. The emerging paradigm, as I have endeavored to describe it, provides us with a far wider and more sophisticated framework for synthesizing the varieties of scientific experience than does the mechanistic and reductionistic one we inherited from the sixteenth and seventeenth centuries. It refines and extends the pluralism natural to science and adds new elements of substance (and not merely of subtlety) to the idea of dialectic. Indeed, it seems to offer, on the whole, a far richer and more adequate point of view for the comprehension of human experience than does the paradigm it is replacing.

REFERENCES

Boring, E. G. (1950) *A history of experimental psychology* (2nd ed.). New York: Appleton-Century-Crofts.

Capek, M. (1961). *The philosophical impact of contemporary physics.* Princeton: Van Nostrand.

Capra, F. (1982). *The turning point: Science, society, and the rising culture.* New York: Simon & Schuster.

Chorover, S. L. (1973). Big brother and psychotechnology. *Psychology Today, 7,* 43–54.

Chorover, S. L. (1974). Psychosurgery: A neuropsychological perspective. *Boston University Law Review, 54,* 231–248.

Chorover, S. L. (1976a). An experimental critique of "consolidation" studies and an alternative "model systems" approach to the psychobiology of learning and memory. In M. R. Rosenzweig & E. L. Bennett (Eds.), *Neural mechanisms of learning and memory. (pp. 561–582).* Cambridge: MIT Press.

Chorover, S. L. (1976b). The pacification of the brain: From phrenology to psychosurgery. In T. P. Morley (Ed.), *Current contoversies in neurosurgery* (pp. 730–767). Philadelphia: Saunders.

Chorover, S. L. (1979). From genesis to genocide: The meaning of human nature and the power of behavior control. Cambridge: MIT Press.

Chorover, S. L. (1980a). The psychosurgery evaluation studies and their impact on the Commission's report. In E. Valenstein (Ed.), The psychosurgery debate: Scientific, legal and ethical perspectives (pp. 245–263). San Francisco: W. H. Freeman.

Chorover, S. L. (1980b). Violence: A localizable problem? In E. Valenstein (Ed.), The psychosurgery debate: Scientific, legal and ethical perspectives (pp. 334–347). San Francisco: W. H. Freeman.

Chorover, S. L. (1985). Psychology in cultural context: The division of labor and the fragmentation of experience. In S. Koch & D. E. Leary (Eds.), *A century of psychology as science* (pp. 870–879). New York: McGraw-Hill.

Chorover, S. L., & Chorover, B. (1982). Towards a theory of human systems. In S. P. R. Rose (Ed.), *Towards a liberatory biology: Proceedings of the Bressanone conference on the dialectics of biology and society in the production of mind (Vol. 2,)*. London: Allison & Busby.

Chorover, S. L., & Cole, M. (1961). An investigation of the equivalency between auditory and visual rhythms in man. Journal of Physiology (London), *156*, 35–36.

Chorover, S. L., & Deluca, A.-M. (1975). Electrically-induced epileptogenesis in rabbit olfactory bulb: Electrophysiological plasticity in a simple cortical system. *Brain Research, 94*, 185–197.

Chorover, S. L., & Schiller, P. H. (1965). Short-term retrograde amnesia in rats. *Journal of Comparative Physiological Psychology 61*, 34–41.

Descartes, R. (1955). *Philosophical works* (Vol. 1). E. S. Haldane & G. R. T. Hall (Eds.). New York: Dover.

Easlea, B. (1981). *Science and sexual oppression*. London: Weidenfeld & Nicolson.

Edelman, P. D. (1978). *Physics and psychology: From the Newtonian to the modern worldview*. S. B. thesis (unpublished), Massachusetts Institute of Technology.

Einstein, A. (1905). Ueber einen die Erzeugung und Verwandlung des Lichtes betroffenden heuristischen Gesichtspunkt. *Annelen Der Physik.*, Ser. 4, *117*, 132–148.

Gergen, K. J. (1985). The social constructionist movement in modern psychology. *American Psychologist, 40*, 266–275.

Griffin S. (1980). *Woman and nature*. New York: Harper/Colophon.

Hanson, N. R. (1958). *Patterns of discovery*. London: Cambridge University Press.

Heisenberg, W. (1958). *Physics and Philosophy*. New York: Harper Torchbooks.

Heisenberg, W. (1972). The representation of nature in contemporary physics. In S. Sears & G. Lord (Eds.), *The discontinuous universe* (pp. 122–135). New York: Basic Books.

Howard, G. S. (1985). The role of values in the science of psychology. *American Psychologist, 40*, 255–265.

James, W. (1950). *Principles of psychology* (2 Vols). New York: Dover. (Original work published 1890)

Kling, A. (1972). Effects of amygdalectomy of serial-affective behavior in non-human primates. In B. E. Eleftheriou (Ed.), *The neurobiology of the amygdala* (pp. 511–536). New York, Plenum Press.

Klüver, H., & Bucy, P. (1939). Preliminary analysis of functions of the temporal lobes in monkeys. *Archives of Neurological Psychiatry, 42*, 979–1000.

Koch, S., & Leary, D. E. (1985). *A century of psychology as science*. New York: McGraw-Hill.

Krasner, L., & Houts, A. C. (1984). A study of the "value" systems of behavioral scientists. *American Psychologist, 39*, 840–850.

Kuhn, T. S. (1962). *The structure of scientific revolutions*. Chicago: University of Chicago Press.

Lasswitz, K. (1890). *Geschichte der Atomistik von Mittelalter bis Newton* (2 Vols). Hamburg: Voss.

Lazarus, R. S. (1984). On the primacy of cognition. *American Psychologist, 39*, 124–129.

Lehrman, D. S. (1953). A critique of Lorenz's "objectivistic" theory of animal behavior. *Quarterly Review of Biology*, 337–363.

Lehrman, D. S. (1962). Interaction of hormonal and experiential influences on development of behavior. In E. L. Bliss (Ed.), *Roots of behavior*. New York: Harper & Row.

Lehrman, D. S. (1964). Control of behavior cycles in reproduction. In W. Etkin (Ed.), *Social behavior and organization among vertebrates*. Chicago: University of Chicago Press.

Liess, W. (1972). *The domination of nature*. New York: Braziller.

Maier, N. R. F., & Schneirla, T. C. (1935) *Principles of animal behavior*. New York: McGraw-Hill.

Mark, V. H., & Ervin, F. R. (1968). Is there a need to evaluate in individuals producing human violence? *Psychiatric Opinion, 8*, 32–34.

Mark, V. H., & Ervin, F. R. (1970). *Violence and the brain*. New York: Harper & Row.

Mark, V. H., Sweet, W. H., & Ervin, F. R. (1967). Brain disease and urban riots. *Journal of the American Medical Association, 201*, 895.

Mendelsohn, E. (1977). The social construction of scientific knowledge. In E. Mendelsohn &

P. Weingert (Eds.), *The social production of scientific knowledge* (pp. 3–26). Dordrecht, Holland: D. Reidel.

Mendelsohn, E., & P. Weingert (1977). *The social production of scientific knowledge*. Dordrecht, Holland: D. Reidel.

Merchant, C. (1980). *The death of nature: Women, ecology, and the scientific revolution*. San Francisco: Harper & Row.

Oppenheimer, J. R. (1950). The age of science: 1900–1950. *Scientific American, 183*, 20–23.

Peele, S. (1981). Reductionism in the psychology of the eighties. *American Psychologist, 36*, 807–818.

Rosvold, H. E., Mirsky, A. F., & Pribram, K. (1954). Influence of amygdalectomy on social behavior in monkeys. *Journal of Comparative and Physiological Psychology, 47*, 173–178.

Sahtouris, E. (1989). *Gaia: The Human journey from chaos to cosmos*. New York: Pocket Books.

Schneirla, T. C. (1949). Levels in the psychological capacities of animals. In R. W. Sellars, V. J. McGill, & M. Farber (Eds.), *Philosophy for the future*. (pp. 243–286). New York: MacMillan.

Schneirla, T. C. (1954). Interrelationships of the "innate" and the "acquired" in instinctive behavior. In P.-P. Grasse (Ed.), *L'instinct dans le comportement des animaux et de l'homme* (pp. 387–452). Paris: Masson.

Schneirla, T. C. (1957). The concept of development in comparative psychology. In D. B. Harris (Ed.), *The concept of development* (pp. 78–108). Minneapolis: University of Minnesota Press.

Skinner, B. F. (1938). *The behavior of organisms*. New York: Appleton-Century.

Sweet, W. H. (1970). Foreword. In V. H. Mark & F. R. Ervin, *Violence and the brain*. New York: Harper & Row.

Toulmin, S. (1982). *The return to cosmology: Postmodern science and the theology of nature*. Berkeley: University of California Press.

Valenstein, E. (1973). *Brain control: A critical examination of brainstimulation and psychosurgery*. New York: Wiley.

Valenstein, E. (1980). *The phychosurgery debate: Scientific, legal, and ethical perspectives*. San Francisco: W. H. Freeman.

Wapner, S. (1983). Living with radical disruptions of person-in-environment systems. *International Association of Traffic & Safety Sciences Review 2*, 133–148 (Japan).

Wapner, S. (1987). A holistic, developmental, systems-oriented environmental psychology: Some beginnings. In D. Stokols & I. Altman (eds.), *Handbook of environmental psychology* (pp. 1433–1465). New York: Wiley.

Watson, J. (1914). *Behavior: An introduction to comparative psychology*. New York: Holt.

Wheeler, J. A. (1982). Bohr, Einstein and the strange lesson of the quantum. In R. Q. Elvee (Ed.), *Mind in Nature* (pp. 1–30). San Francisco: Harper & Row.

Zajonc, R. B. (1984). On the primacy of affect. *American Psychologist, 39*, 117–123.

7 A Sociocultural Approach to the Study of Mind

Sylvia Scribner
The Graduate School and University Center
City University of New York

This chapter examines the contribution of the sociohistorical school of psychology to the construction of a scientific methodology of mind. What are the philosophical premises and guiding constructs of this school, and how does it frame research questions about the origin and development of mind? After a brief theoretical exposition, I describe and analyze a series of studies designed to test a particular hypothesis about cognitive change derived from this theory. In this account, I "descend" to a level of description involving particulars of research design and interpretation not often included in a treatment of epistemological issues. I have two justifications for this approach. First, a scientific epistemology, as distinct from one based on metaphysical speculation, is grounded on knowledge gained through the empirical procedures of the various disciplines investigating mind. We need to concern ourselves, then, with the logic, possibilities, and limitations of the research methods now on hand in these disciplines and the questions they most suitably address. A second reason is pragmatic. Exemplifying a theoretical approach enhances its graspability. Because the social-historical school of psychology is only now becoming widely known, I want to bring both example and principle to bear in considering its contribution to the epistemic enterprise.

One other introductory observation is in order. As our discussion here demonstrates, the terms *mind*, *cognition*, *knowing*, and *knowledge* tend to acquire different content as they move from one discourse or scientific frame of reference to another. Or, more troublesome still, they seem at times to lose concrete content altogether, blending into general abstract concepts that are not amenable to any particular methodology or disciplinary attack. Let me put my enterprise into perspective then, by taking the general concept of cognition and giving it specific content as an object of inquiry in theory and research.

As a cognitive developmental psychologist, my interest is centered on human cognition, and my research is addressed to questions about the structure, function, and mode of development of human cognition. More specifically, I have been concerned with understanding those aspects of human cognition that early psychologists (Wundt, 1916, for example, and see discussion on Vygotsky following) referred to as the "higher psychological functions." They subsumed under this term such complex processes as logical reasoning, intentional memory, planning, and decision making, among others. Although the term has fallen out of favor, the new cognitive sciences, especially those employing information-processing and artificial intelligence models, (e.g., Newell & Simon, 1972, on problem solving; Johnson-Laird, 1983, on syllogistic reasoning; Gentner & Gentner, 1983, on mental models, among others) and much of developmental psychology (Brown, 1978; Brown, Bransford, Ferrara & Campione, 1983, on metacognition; Piaget & Inhelder, 1958, on propositional reasoning; Joffe-Falmagne, 1980, on deductive inference) continue to devote major resources to conceptualizing and studying these complex mental phenomena.

Tobach (1987) holds that a scientific understanding of changing (developing) phenomena rests on an identification of both continuities and discontinuities in underlying processes. This principle would certainly seem to carry special force in an inquiry into the grounds of knowing. However, as a number of contributions to this volume attest, the deployment of neo-Darwinian concepts and mechanisms to epistemological objects maintains a one-sided focus on cross-species continuities in cognition. Missing from theories involving biological mechanisms (e.g., blind variation; Campbell, 1960) as primary means of knowledge acquisition throughout the phylogenetic scale is any considered treatment of discontinuities—in particular, the special modes of knowing that arise in human society and bring into being those qualitatively distinct, "higher" intellectual functions that generations of psychologists have sought to understand and that remain the central concern of philosophical epistemology. It is the leading achievement of the sociohistorical school of psychology that it puts forward an account of these functions that resorts to neither vitalism nor reductionism (Vygotsky, 1981); this account, therefore, offers important theoretical resources for the construction of a scientific epistemology.

SOCIOHISTORICAL THEORY AND COGNITIVE PSYCHOLOGY

L.S. Vygotsky (1896–1934) is the founder of a dialectical materialist approach to psychological theory that has been variously characterized as a "sociocultural" or "sociohistorical" framework. His students and colleagues A.R. Luria and

A.N. Leont'ev are among the better known—together with other psychologists and philosophers in the Soviet Union elaborated and systematized this framework, which, in recent years, has become internationalized (Engestrom, 1987; Hildebrand-Nilshon & Ruckriem, 1988; Wertsch, 1981). Now identified as "activity theory," this project may be more accurately characterized on the level of *meta*theory. Seeking to overcome Cartesian dualities, it lays out new analytic categories for the construction of domain-level theories within the respective subfields of epistemology (Lektorsky, 1980) and psychology (Leont'ev, 1978, 1979). Within psychology, various schools of thought (Brushlinskii, 1987; Kozulin, 1986; Laboratory for Comparative Human Cognition, 1983; Wertsch, 1985) offer alternative views of how these basic level categories can best be brought to bear in investigating and explaining processes of learning and development. In spite of these divergences, most scholars accept Vygotsky's (1978; 1986) main ideas as essential building blocks.

I present here (Fig. 7.1) a simplified and schematic sketch of these building blocks as they bear on the topic at hand: human cognition.[1]

Three terms capture the principal features distinguishing human knowing from that of other species: Human cognition is *culturally mediated*; it is founded on *purposive activity*; and it is *historically developing*. These are three interconnected constructs, and although I consider them individually, we will see how they work together.

A defining characteristic of human behavior generally and of human cognition in particular that sets it apart from the behavior of other species is that humans do not interact directly with the environment (Vygotsky, 1978; 1981). Rather, they regulate their interactions with each other and with the environment through systems of "objects." These objects stand in-between, or "mediate" human actions in the world. The classic case of human activity mediated by objects is labor; and Marx's (1967) description of the role of tools or instruments in this process served Vygotsky as a prototype of all human actions: "The elementary factors of the labour-process are 1) the personal activity of man, i.e., work itself, 2) the subject of that work, and 3) its instruments. . . . An instrument of labour is a thing, or a complex of things, which the labourer interposes between himself and the subject of his labour, and which serves as the conductor of his activity" (p. 174).

Vygotsky extended this notion of mediating instruments to human intellectual as well as behavioral functions. Human thinking, he said, is mediated by signs, just as human behavior in the world—practical behavior—is mediated by tools.

[1]Leont'ev (1981) provides a comprehensive and creative account of the origin and development of mind across the entire phylogentic spectrum from an activity theory perspective. His account, with contributions dating from the early 1940s, serves as an early model of an empirically based epistemology.

BASIC ATTRIBUTES OF HUMAN COGNITION
(Higher Psychological Functions)

STRUCTURE	Mediated (tools/signs)
FUNCTION	In goal-directed activities
MODE OF DEVELOPMENT	Historical/Social

FIG. 7.1 Sociohistorical Theory of Mind.

Incorporation of cultural sign systems into human mental activity is the decisive moment in evolutionary history, transforming "elementary" cognitive processes into the complex functional systems ("higher" functions) that characterize mature thought in all human societies. (See Leont'ev & Luria, 1968, for later criticism of this view.)

A rudimentary illustration of sign-mediated cognition is remembering with the aid of external cues. The practice of cutting notches in sticks is an early example of human use of objects as "conductors" of mental activity. Instead of relying on the spontaneous appearance of cues for recall, humans introduce changes in the world that help them regulate and control their own memory processes. The notched stick has no significance as a physical object; it is what the notch stands for—its meaning—that affects the remembering process. In human consciousness, significant mediating "signs" are not simply isolated things of this kind, but complex symbolic systems. Language is the principal and universal sign system, which functions not only as the prime vehicle of species communication but as the means through which humans acquire conceptual thought. Significant mediators include number, writing, and other notational systems, and more broadly, all artifactual and ideational (knowledge, theories) systems through which and by means of which humans cognize the world. As is clear, "symbols" or signs in this theory are not the meaningless strings of manipulable elements that feature in computer models of mind; they are semantic objects. Mediators are cultural, therefore, in the most profound sense: All meaning-systems whether in the form of beliefs, knowledge, or scientific theories are products of human culture; and they are invented and transmitted through the social processes and institutions by which succeeding generations reproduce and change their cultural heritage.

The concept that human activity, including its cognitive aspects, has a tripartite structure has profound theoretical implications. Psychological theories have historically attempted to account for behavioral and mental development through two terms: *organism* and *environment*. Theories are distinguished from one another by the way in which they combine these terms; "environmental determinists" stress the primacy of one term; "nativists" or "organicists" stress the primacy of the other; and "interactionists" stress the reciprocal effects of both. A tripartite scheme—*organism*, *environment*, and *cultural mediator*—radically changes the structure of explanation of human behavior and cognition (Davydov & Radzikhovsky, 1985), and provides a new foundation for epistemological endeavors. (Lektorsky, 1980, develops these implications from a philosophical standpoint.)

We move now to the second term: *purposive activity*. In the view of the sociohistorical school, mental and behavioral processes are aspects of a unitary system of human action-in-the-world. Planning, remembering, reasoning, and problem solving—cognitive acts in general—are not undertaken as isolated acts or ends in themselves; they are not functions of a mind cut off from a body. Rather they are indissolubly tied to behavioral actions, with both embedded in larger systems of activity whose purposes they serve. Cognition functions as an integral aspect of purposive activities. In every human society, socially patterned systems of activities (sometimes referred to as social practices) arise to satisfy human needs and aspirations: We play, we work, we praise the gods, we conduct experiments, we go to school. These larger, motivationally driven activities provide what is loosely called the ''context'' of cognition. Such activities may be primarily directed toward practical ends (productive labor is the prototype here) or they may serve aesthetic, ideational or affective ends; all are directed at satisfying motives that are socially generated, and all are carried out through social means and modes of action. As individuals participate in these socially patterned activities, they acquire the cultural forms of acting and thinking incorporated in them; reciprocally, they contribute to changing these activities and to inventing new modes of action and thought. In the contemporary form of the sociohistorical school (activity theory), this category of purposive activity has become the central explanatory construct for understanding behavioral and mental development. For cognitive inquiry, the theoretical implication is that empirical sciences such as developmental psychology need to shift away from preoccupation with the ''individual human mind'' as the sole or principal unit of analysis for understanding modes of thought (as in Piagetian theory) to a consideration of the systems of socially constituted activity in which human consciousness and behavior develop.

Finally we come to *mode of development*. It follows from the culturally mediated nature of human psychological systems and their functional role in social practices that changes in the sociocultural level of organization will have major implications for changes in psychological organization. Social practices and activities change over time, and so do symbolic mediational means. Such transformations are to be understood as constructed historically through human action, rather than resulting from changes in the biological constitution of the human organism. Although evolutionary mechanisms account for the origins of elementary mental processes, their organization into functional, sign-mediated systems comes about through historical processes: History, rather than phylogeny, becomes the dominant motor of cognitive change. Vygotsky expressed this conclusion in the following way(cited in Scribner, 1985): "Human behavior differs from animal behavior in the same qualitative manner as the entire type of adaptability and historical development of man differs from the adaptability and development of animals, because the process of man's *mental development* is part of *the general historic development of mankind* [emphasis added]" (p. 123).

Note that this view accounts for both universal and culturally specific features of human thought. All humans participate in a general species history in which human activities became organized around symbolic-communicative and technical means; thus all exhibit "higher mental functions." But because activities and means vary over time and space, these cognitive functional systems will also reflect such differences. (For a fuller exposition of Vygotsky's historical methodology, see Scribner, 1985.)

("Evolutionary epistemology" in this view serves as only one leg on which a scientific methodology of mind can stand; "historical epistemology" is the other, Wartofsky, 1982).

The distinctive contribution of sociohistorical theory of cognition is also highlighted when it is contrasted to another contemporary epistemologic enterprise that seeks empirical grounding—the effort by certain analytic philosophers to formalize relations between philosophy's classic concern with the grounds for "true and justified belief" and the knowledge psychology provides about cognitive processes. Goldman's (1986) attempt at synthesis is of special interest here, for he makes clear that he looks to cognitive science to provide knowledge about the *human* mind because the human mind's "belief-forming processes" are relevant to epistemology. Goldman's human mind, however, is timeless. It has an "architecture" or "intrinsic properties" that Goldman looks to psychology to identify and describe; epistemology is concern with "native cognitive mechanisms" whose strength and weaknesses epistemology assesses. Thus, on the other hand, we confront theories (e.g., Campbell) that focus on evolutionary mechanisms of change in mental processes but that fail to account for the special characteristics of human mind; on the other, we have theories (e.g., Goldman) that take the distinctive characteristics of human mind as the object of inquiry but that fail to account for its origin in evolution or change over historical time. Vygotsky's sociohistorical theory contributes a perspective for overcoming this opposition.

Methodology

Certain methodological principles are implicated in this approach. The theory suggests that changes on the sociocultural level—changes in activities and mediational means—will afford possibilities for changes in individual cognition. Accordingly, an analysis of changing social practices becomes integral to—rather than merely peripheral to—an inquiry into learning and development. This framework also places a premium on the comparative method, one that searches for commonalities and differences among social practices within or between cultures. In certain domains of investigation, the accumulated research base and conceptual framework may be adequate for the formulation of specific hypotheses about the way in which changes in social activities articulate with cognitive change. However, as is the case with other "grand" or metatheories, activity theory provides no formula for a search for specific mediating mechanisms. Working from the level of general theory to the generation of particular hypotheses requires the usual mix of knowledge, example, and intuition. Fortunately, examples of useful empirical investigations within the sociohistorical theoretical approach are accumulating (Engestrom & Engestrom, 1985; Hacker, 1982; Saxe, 1982; Scribner, 1984; Tulviste, 1979), providing grist for further theory development.

I describe one such investigation, drawing on my own experience, to illustrate both the problems and possibilities of this approach. The research question stemmed directly from Vygotsky's hypothesis about symbolic mediators and their cognitive implications: Specifically, we attempted to test empirically whether mastery of a written language systems fosters special modes of memory and thinking.

Speculations on Literacy and Thought

The choice of literacy as a domain for testing a sociocultural approach to cognition had certain obvious advantages. We cannot doubt that writing systems are a human invention whose origins lie in history rather than in biological evolution. Characteristics and uses of earliest known scripts have been shown to change over time in consonance with changing social practices—expansion of trade, for example (Schmandt-Besserat, 1978) or changes in modes or religious education (Resnick & Resnick, 1988). And, a great deal of evidence, drawn from various historical epochs, indicates that the introduction of writing systems often went hand-in-hand with the accumulation of knowledge and changes in intellectual practices (Goody, 1968). Most challenging was the fact that, although speculations about the relationship between "writing" and "mind" stretch back to antiquity, great thinkers failed to reach consensus about the nature of that relationship. Contrary responses of Greek philosophers to the spread of literacy among them characterize this age-old debate. Rational analysis suggested to Socrates that writing would diminish human powers by leading to forgetfulness and a propensity to confuse "words" and "ideas." Rational analysis suggested to Plato, however, that writing would strengthen human reason by freeing it from the emotional connotations that inhere in language in the oral mode.

Since the last century, comparative scholars in such fields as history, anthropology, and linguistics have sought to bring philosophical speculation into contact with evidence derived from field work studies and documentary analyses. Goody (1977, 1987), a leading scholar in this field, has concluded that the weight of this evidence strongly supports an "enhancement" position: Literacy (especially alphabetic literacy) fosters abstract reasoning, logical inference, hierarchical classification, and other higher cognitive functions.

Significant as these analyses are for informing us of the intricate pattern of relationships among literacy activities and other cultural practices, they fall short of providing critical evidence for generalizations about cognitive change in individuals. Social science scholars have assumed that these relationships—for example, the relationships between alphabetic literacy in Greece and Aristotle's invention of the syllogism—are mediated by changed modes of thinking in individuals, but they offer no direct evidence that this is the case. Historically, the kinds of observations that substantiate claims about individual thought processes have been the province of psychology and a product of its theories and investigative methods. To support the thesis that literacy makes a difference in mental processes, psychological analysis has to be joined with cultural analysis. Vygotsky recognized this necessity, and it is not accidental that he was the first to organize *experimental* research into the consequences of schooling and literacy (Luria, 1976; for an appraisal of this research, see Cole, 1988).

A project for identifying literacy as a mediator of cognitive change puts stringent demands on research design. Prospective longitudinal studies require more control over literacy education programs for a longer period of time with a larger group of people than most researchers can achieve. Cross-sectional comparisons of literate and nonliterate individuals appear more feasible but these must meet many constraints:

- To sustain inferences that literacy is the crucial mediating mechanism requires comparisons among literate and nonliterate people living in the same social milieu, speaking the same language, and experiencing life circumstances that do not differ in any systematic way. Precluded, therefore, are such ready-to-hand comparisons as those between affluent literates and poor nonliterates, or between literates who speak and write English and nonliterates who speak Spanish.
- It requires that literacy be disentangled from other educative experiences such as knowledge acquisition, which co-occur with literacy instruction in school.
- Finally, it requires that we distinguish cognitive changes that may be associated with literacy from general developmental changes in intellectual function, such as those posited by Piaget for the period up to and including adolescence. Precluded, therefore, is any research design in which increasing mastery of literacy covaries with increasing age.

Such constraints argue for a within-culture comparison of "like" adults among whom some form of literacy is achieved outside of schooling—a prescriptive model that forces researchers to seek specialized sites for their inquiry and forces psychologists out of the laboratory into the field.

Literacy in Vai Land

We were fortunate in finding a site among the Vai people in West Africa that satisfied many of these constraints. The Vai are a Mande-speaking people who, like their neighbors, practice slash-and-burn rice farming using simple iron tools, but they have attained a special place in world history as one of the few cultures to have independently invented a phonetic writing system (Dalby, 1967; Gelb, 1952; Koelle, 1854). The Vai script is a syllabary of some 200 characters and has been in active use within the context of traditional rural life for almost 2 centuries. Widely available to all members of the society (although in practice restricted to men), Vai script is passed on from one individual to another, much as are other specialized skills such as weaving or tailoring; schooling is not involved nor is there any institutionalized profession for teaching the script. Moreover, Vai script coexists with two universalistic and institutionally powerful scripts—the Arabic and English alphabets—which are also used by some Vai people in the village setting. Arabic literacy is acquired through instruction in the Qu'ran (relying primarily on recitation and memorization), and English literacy is acquired through attendance in Western-type schools. These circumstances established conditions for a "natural experiment," as it were, in which we could compare unschooled literacies (Vai script and Arabic) with literacy-cum-schooling (English), and the two nonschooled literacies with each other. (For a full description of the Vai literacy research program, see Scribner & Cole, 1981.)

An Interdisciplinary Approach

We assembled an interdisciplinary team including anthropologists (one of whom was a Vai man), experts in the Vai script, and cognitive psychologists. The first phase of the research consisted of a sociocultural analysis of literacy practices in each of the scripts. Through ethnographic observations and systematic survey interviews, we sought to determine the larger activities in which reading and writing tasks were embedded. How were the scripts used in ritual, trade, and craft, in family relations, in government administration? We sought, too, to outline principal characteristics of the instructional settings in which each script was learned, a project that led us to conduct on-site field studies of Qu'ranic learning in the villages and to document various forms of tutorial instruction in the Vai script. Not surprisingly, we found great diversity in practices associated with each script: Each literacy had its own distinguishing features, mode of instruction, specialized social functions, cultural meaning, and symbolic significance. There was no "literacy in general" among the Vai.

We were now in a position to undertake psychological studies focusing on literacy as a mechanisms of cognitive change. In our first experimental assay, we used standard experimental paradigms that were initially designed to serve as indices of developmental status among Western-schooled children and adults, but which had also been widely used in cross-cultural research. These included classification and word memory tasks, syllogistic reasoning problems, similarity judgments, and the like. A principal outcome of these studies was a "school effect": Only those Vai people who had attended Western-type schools consistently performed in accordance with Western norms. Because schooled people were also the only participants with prior practice in tasks of this kind (most standard experimental tasks involve skills implicated in school performance), this outcome was basically uninterpretable as well as unilluminating with respect to our literacy hypotheses.

With this experience on hand, we reevaluated the methodological implications of our theoretical framework. This framework emphasizes the specificity of mediational systems and the varying as well as the common aspects of social practices. It appears that, if we want to use experiments to investigate the relationship between some aspect of social practice and cognition, we need to construct tasks that bear a closer relationship to the practices of interest than standard laboratory pradigms. We turned again to observational procedures, this time for a more fine-grained analysis of the modal reading and writing tasks required in literacy practices involving the different scripts. After identifying such tasks, we analyzed their component cognitive skills and devised experimental tasks embodying them. Rather than function as general instruments of assessment, our tasks served as models of those arising in naturally occurring activities. We used them to test a generalization question: Did the skills we had identified in everyday literacy activities generalize to cognitive tasks encountered outside the context of the original practice? Here by way of illustration is a description of one line of investigation involving hypothesized skills in Vai script literacy.

Illustrative Experiment

Reading hand-written letters from friends and kin was one of the most common functions of Vai script literacy, and our observations indicated that it was a cognitively demanding one. Graphic signs in the Vai script represent syllables in the spoken language, but there is not a perfect correspondence between them. Vowel tone and vowel length, two semantically crucial features for understanding spoken Vai, are not represented in the script and need to be inferred from context. Writing conventions, as in ancient Greece, do not segment lines of text into words or other semantic or syntactic units. Each syllable stands alone with space surrounding it. Because Vai words may consist of more than one syllable, readers must find a way of integrating individual syllables on the sheet of paper into meaningful chunks of language. A common solution to these problems is reading aloud—grouping and regrouping strings of syllables that are problematic and varying their pronounciation until they are organized into such meaningful chunks. In this recycling process, readers must keep separate syllables in mind until they can be integrated into words or phrases. We supposed that this experience might foster skills in integration of linguistic units, and that these skills might be available for language processing in contexts that did not involve the script.

To test this idea, we developed an oral comprehension and recall task. We asked nonliterates and members of the various literacy groups to listen to tape recordings of a native Vai speaker. The speaker presented meaningful Vai sentences one at a time without tonal variations, mimicing their representation in the script. Each informant received two sets of sentences; sentences in one set were segmented into nine syllable units separated by a 2-second pause; sentences in the other set were segmented into an equal number of word units. Word unit sentences served as a baseline control for possible individual or group differences in hearing or memory. Informants were asked to repeat each sentence immediately after hearing it, and to answer a comprehension question about it.

We conducted a number of studies with various population groups, replicating basic results. Table 7.1 displays the general pattern of our findings—namely, a striking interaction between population group and task materials.

There were no distinguishable differences among literacy groups in recall/comprehension (''perfect recall'') of sentences segmented into words, and all literacy groups were superior in performance to nonliterates. But recall dropped precipitously on syllable unit sentences for all groups except those classified as Advanced Vai script readers based on reading test scores. Proficient Vai script readers had comparable performances on words and syllables, whereas readers of other scripts dropped to the level of nonliterates when syllables were substituted for words. Regression analysis confirmed the contribution of Vai script reading score to syllable recall measures. We performed an even more stringent test: We entered each person's recall score for word sentences as a control in equations predicting syllable recall scores. We wanted to determine whether additional factors, over and above those contributing to word sentence recall, facilitated integration and interpretation of syllables. Of 20 background factors examined, only one affected performance—advanced reading scores in Vai script. This

TABLE 7.1
Means Scores on Auditory Task, Integration Survey

Literacy Group	*Word Sentences Perfectly Recalled (max 5)*	*Syllable Sentences Perfectly Recalled (max 5)*
Nonliterate men	1.45	.90
Arabic literates		
Qur'anic only	2.50	1.00
Arabic language	2.60	1.10
Beginning Vai script literates	2.80	1.10
Advanced Vai script literates	2.85	2.40

Note: Adapted from Scribner & Cole (1981).

outcome provides the most compelling evidence that ability to understand, remember, and reproduce sentences parsed into syllables reflects expertise in reading Vai script—no other literacy or experimental factors examined in our analyses offered an alternative route to these skills. Moreover, these analyses indicated that it was not merely exposure to Vai script literacy but level of mastery attained that was critical to performance; the fact that different levels of proficiency within a single literacy group made a difference strengthens the argument that the skills involved in the experimental tasks were indeed literacy related. We went one step further in unpackaging "Advanced Vai Script" proficiency. We compared individuals classified in the Beginning and Advanced Vai script groups and found them similar in demographic characteristics such as age, knowledge of languages, and the like but quite different in the quantity and quality of their literacy activities. Advanced Vai script informants had known the script twice as long as Beginning Vai informants; they reported using it more often and for more diversified purposes; all had taught the script to someone else whereas only a minority of Beginning Vai had ever taught it.

These experimental studies suggest that practice in reading may promote the acquisition of specialized skills for linguistic analysis in the auditory mode: Reading may improve listening comprehension. This result is neither trivial nor self-evident. Recall that all Vai people in these studies were native speakers; no group had more or less experience listening to naturally spoken Vai than any other. Vai script readers, however, were also required to hear and make sense of segmented Vai, a form of linguistic analysis not demanded in situations of oral communication. Because our observational and experimental evidence implicate the same set of skills, we can present a strong case for the formative role of literacy practice.

In other experiments designed on the same lines, we demonstrated literacy effects on higher order cognitive skills. Vai script literates engage in discussion and criticism of modes of expression in written communication—a practice that enhanced their ability to detect and explain syntactical deviations in spoken Vai. These literacies, with extensive experience in letter writing, were also superior to others in a variety of oral instructional tasks that required the transmission of explicit information in a well-structured form. As for Arabic literates, we found that a memory strategy used for rote learning of Qu'ranic verses was generalized to an experimental recall tasks in which it was an efficient means of accurate information retrieval. Moreover, the strategy was more effectively used by those with longer periods of Qu'ranic study than by novice students.

The skills associated with literacy in Vai country are not necessarily associated with everyday literacy in other societies, nor do literacy activities here represent the full range of practical and intellectual functions of literacy in other parts of the world. The general significance of the research, therefore, lies not so much in its specific findings as in the methodology it developed for analyzing how changes in symbolic mediators (e.g., new writing systems) and in social practices (e.g., communication by letter instead of messenger) affect changes in individual cognition. The methodology is time consuming, labor intensive, and dependent on the organization of knowledge and techniques from a number of disciplines. The information it yields is detailed and local, and it speaks to theory in a cumulative fashion, an incremental manner; it neither upholds nor disconfirms grand theories. But we hope that small gains made by many investigators will, in time, contribute to a more systematic theory of the sociohistorical development of human thought and the working out of an integrative framework in which human thought takes its place in the evolutionary scheme.

REFERENCES

Brown, A. L. (1978). Knowing when, where and how to remember: A problem of metacognition. In R. Glaser (Ed.), *Advances in instructional psychology* (Vol. 1). Hillsdale, NJ: Lawrence Erlbaum Associates.

Brown, A. L., Bransford, J., Ferrara, R., & Campione, J. (1983). Learning, remembering and understanding. In P. H. Mussen (Ed.), *Handbook of child psychology* (Vol. I). New York: Wiley.

Brushlinskii, A. V. (1987). Activity, action and mind as process. *Soviet Psychology*, *25*, 59–81.

Campbell, D. T. (1960). Blind variation and selective retention in creative thought as in other knowledge processes. *Psychological Review*, *67*, 380–400.

Cole, M. (1988). Cross-cultural research in the sociohistorical tradition. In M. Hildebrand-Nilshon & George Ruckriem (Eds.), *Proceedings of the 1st international congress on activity theory* (Vol. 1). Berlin: Druck & Verlag System Druck, 65–86.

Dalby, D. (1967). Further indigenous scripts of West Africa: Mandingo, Wolof and Fula alphabets and Yoruba "holy" writing. *African Language Studies*, *8*, 1–51.

Davydov, V. V., & Radzikhovsky, L. A. (1985). Vygotsky's theory and the activity-oriented approach in psychology. In J. V. Wertsch (Ed.), *Culture, communication and cognition*. New York: Cambridge University Press.

Engestrom, Y. (1987). *Learning key expanding*. Helskinki: Poinette Gummerus Org.

Engestrom, Y., & Engestrom, R. (1985). Developmental work research: The approach and an application in cleaning work. *Nordisk Pedagogik*, *6*, 2–15.

Gelb, I. J. (1952). *A study of writing*. Chicago: The University of Chicago Press.

Gentner, D., & Gentner, A. (Eds.). (1983). *Mental models*. Hillsdale, NJ: Lawrence Erlbaum Associates.

Goldman, A. I. (1986). *Epistemology and cognition*. Cambridge, MA: Harvard University Press.

Goody, J. (Ed.) (1968). *Literacy in traditional societies*. Cambridge: Cambridge University Press.

Goody, J. (1977). *The domestication of the savage mind*. Cambridge: Cambridge University Press.

Goody, J. (1987). *The interface between the written and the oral*. Cambridge: Cambridge University Press.

Hacker, W. (1982). Objective and subjective organization of work activities. In M. von Cranach & R. Harre (Eds.), *The analysis of action*. New York: Cambridge University Press.

Hildebrand-Nilshon, M., & George Ruckriem (Eds.). (1988). *Proceedings of the 1st international congress on activity theory* (Vol. 1, pp. 65–86). Berlin: Druck & Verlag System Druck.

Joff-Falmagne, R. (1980). The development of logical competence: A psycholinguistic perspective. In R. Kluwve & H. Spada (Eds.), *Developmental models of thinking*. New York: Academic Press.

Johnson-Laird, P. N. (1983). *Mental models*. Cambridge, MA: Harvard University Press.

Koelle, S. W. (1854). *Outlines of a grammar of the Vai language*. London: Church Missionary House.

Kozulin, A. (1986). The concept of activity in Soviet psychology. *American Psychologist, 41*, 264–274.

Laboratory for Comparative Human Cognition (1983). Culture and cognitive development. In P. H. Mussen (Ed.), *Handbook of child psychology* (Vol. I). New York: Wiley.

Lektorsky, V. A. (1980). *Subject object cognition*. Moscow: Progress Publishers.

Leont'ev, A. N. (1978). *Activity, consciousness and personality*. Englewood Cliffs, NJ: Prentice-Hall.

Leont'ev, A. N. (1979). The problem of activity in psychology. In J. V. Wertsch (Ed.), *The concept of activity in Soviet psychology*. White Plains, NY: M. E. Sharpe.

Leont'ev, A. N. (1981). *Problems of the development of mind*. Moscow: Progress Publishers.

Leont'ev, A. N., & Luria, A. R. (1968). The psychological ideas of L. S. Vygotsky. In B. B. Wolman (Ed.), *Historical roots of contemporary psychology*. New York: Harper & Row.

Luria, A. R. (1976). *Cognitive development: Its cultural and social foundations*. Cambridge, MA: Harvard University Press.

Marx, K. (1967). *Capital* (Vol. 1). New York: International Publishers.

Newell, A., & Simon, H. A. (1972). *Human problem-solving*. Englewood Cliffs, NJ: Prentice-Hall.

Piaget, J., & Inhelder, B. (1958). *The growth of logical thinking from childhood to adolescence*. New York: Basic Books.

Resnick, D. P., & Resnick, L. B. (1988). The nature of literacy. In E. R. Kintgen, B. M. Kroll, & M. Rose (Eds.), *Perspectives on literacy*. Carbondale: Southern Illinois University Press.

Saxe, G. B. (1982). Developing forms of arithmetic thought among the Oksapmin of Papua New Guinea. *Developmental Psychology, 18*, 583–594.

Schmandt-Besserat, D. (1978). The earliest precursor of writing. *Scientific American, 238*, 38–47.

Scribner, S. (1984). Studying working intelligence. In B. Rogoff & J. Love (Eds.), *Everyday cognition: Its development in social context*. New York: Cambridge University Press.

Scribner, S. (1985). Vygotsky's uses of history. In J. V. Wertsch (Ed.), *Culture, communication and cognition*. New York: Cambridge University Press.

Scribner, S. & Cole, M. (1981). *The psychology of literacy*. Cambridge, MA: Harvard University Press.

Tobach, E. (1987). Integrative levels in the comparative psychology of cognition, language and consciousness. In G. Greenberg & E Tobach (Eds.), *Cognition, language and consciousness: Integrative levels. Proceedings of the 2nd T. C. Schneirla Conference*. Hillsdale, NJ: Lawrence Erlbaum Associates.

Tulviste, P. (1979). On the origins of theoretic syllogistic reasoning in culture and the child. *The Quarterly Newsletter of the Laboratory of Comparative Human Cognition, 1*, 73–80.

Vygotsky, L. S. (1978). *Mind in society*. M. Cole, V. John-Steiner, S. Scribner, & E. Souberman (Eds.). Cambridge, MA: Harvard University Press.

Vygotsky, L. S. (1981). The genesis of higher mental functions. In J. V. Wertsch (Ed.), *The concept of activity in Soviet psychology*. Armonk, NY: M. E. Sharpe.

Vygotsky, L. S. (1986). *Problems of general psychology. Vol. 1 Thinking and speech*. New York: Plenum Press.

Wartofsky, M. (1982). From genetic epistemology to historical epistemology: Kant, Marx, and Piaget. In L. S. Liben (Ed.), *Piaget and the foundations of knowledge* (pp. 1–17). Hillsdale, NJ: Lawrence Erlbaum Associates.

Wertsch, J. V. (1985). *Vygotsky and the social formation of mind*. Cambridge, MA: Harvard University Press.

Wundt, W. (1916). *Elements of folk psychology*. London: Allen & Unwin.

DISCUSSION

8 A Phylogenetic Epistemology: The Evolution of Sensory Organs, Nervous Systems, and Effectors from Protozoan to Primate

Donald D. Jensen
University of Nebraska–Lincoln

In many contexts it is useful to distinguish *process* and *product*, to distinguish that which creates or produces from that which is created or produced. When considering the topic of evolutionary epistemology, some authors deal with the *process* of evolution. For example, D. T. Campbell (1960) has argued persuasively that the process of biological evolution is basically similar to both animal problem solving and human creative thinking, all three involving unguided search or variation and selective retention or reproduction. This chapter does not attempt to add to or clarify our understanding of the process of evolution; instead it focuses entirely on the *products* of biological evolution, that is, on certain specific animal forms that have been produced by evolutionary processes. This chapter asks what the sequence of forms, the evolutionary levels leading from protozoa to primate, can tell us about evolutionary epistemology.

Biological, Behavioral, and Phylogenetic Levels

To prevent confusion, it is important to identify three different meanings of level, only one of which corresponds to evolutionary complexity and phylogenetic position. Level may refer, first, to biological molarity or molecularity, second, to cognitive or behavioral complexity, and, third, to phylogenetic stage or position in a phylogeny. These three meanings correspond to three different portions of a biotemporal space.

The biotemporal space (Jensen, 1987) is a two-dimensional representation of the conceptual field that includes all biobehavioral events (Fig. 8.1). The vertical axis of the biotemporal space corresponds to the dimension of biological molarity

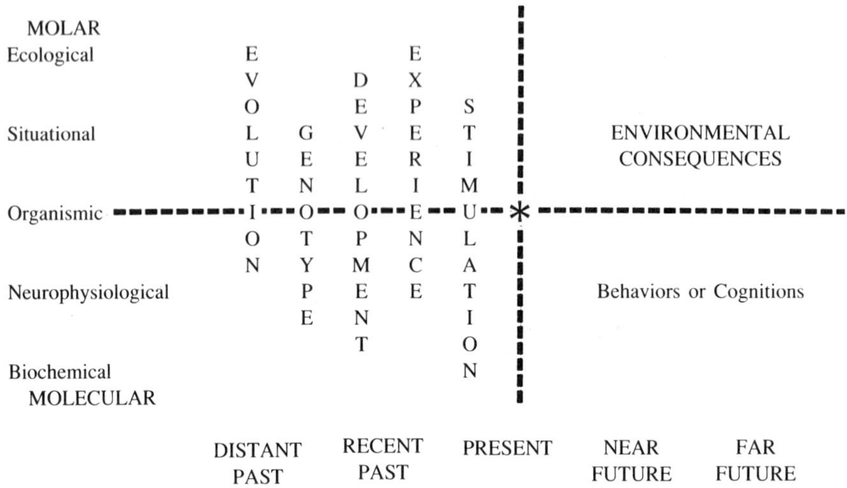

FIG. 8.1 The biotemporal space.

and molecularity—to the different scales at which biological phenomena may be studied (ecological, situational, organismic, neurophysiological, biochemical). Perpendicular to that dimension is the horizontal dimension of time and causation, going from the distant past to the present and to the distant future. Different biobehavioral concepts bind together events located at different points in this field. Behavioral events, located at the present time and at the organismic level, correspond to patterns of organismic events of brief duration; such behavioral events may be influenced or determined by events at the organismic level (e.g., phenotype), by events at more molecular levels (e.g., neurotransmitter availability), by events at more molar levels (e.g., stimuli in the environment, ecology), and by events in the distant past (e.g., embryology, phylogeny), and may themselves influence events in the future at all levels. Evolutionary history, organismic genotype, developmental history, prior experience, and current stimulation correspond to sets of determinants that are arrayed nearer and nearer to the behavioral events at the center of the biotemporal space. Consequences at all levels are represented to the right of the "here and now." More molar events are represented higher and more molecular events are represented lower on the biotemporal space.

Where in Fig. 8.1 are the three kinds of levels? Biological levels constitute the vertical dimension of the biotemporal space. Behavioral or cognitive levels involve recognition of important variation in complexity of different behavioral and cognitive processes, all located at the organismic level and in present time. Phylogenetic levels (and the focus of this chapter) refer to stages in the evolutionary history of the organisms on the human ancestral line and correspond to temporally distant historical events (i.e., earlier organisms ancestral to current organisms, and relevant to their behavior and cognition). Although it is certainly possible to consider the evolutionary development of the individual behavioral and cognitive processes (Schneirla, 1949; Campbell, 1974; Thomas, 1986; Medicus, 1987), that is not the goal of this chapter, which intends only to give the broad features and major characteristics of the phylogenetic development of the human knowledge apparatus.

When discussing evolutionary stages or levels, a disclaimer is necessary. It is clear that evolution has created a bush of many extinct and few living forms, that current forms are not ancestors, and that ancestral forms are not identical to living forms. Nevertheless, it is appropriate to designate earlier stages in a phylogeny by reference to the name of the existing animals with which the ancestor would be classified biologically. Thus, humans did not evolve from any existing ape or monkey, but it is almost certain that we evolved from animals that, if alive today, would be classified with the great apes, rather than any other group, and that these evolved from animals, that, if alive today, would be classified with modern-day monkeys rather than any other biological group, etc. Therefore, ancestral forms will be identified by the names of the existing forms or taxa with which they would be classified, even though they almost certainly differed from existing forms. The ancestral forms are intended to be eotypes, in the terminology of Hanson (1977) (i.e., organisms with primitive features from which modern forms evolved). It is possible to use prefixes and suffixes to designate ancestral forms or eotypes (i.e., protomonkey, lemuroid, nemertoid, archeorhabdocoel, etc.), but it is also possible to abstain from such clumsy expressions and use context and grammar to differentiate between contemporaneous organisms and early ancestors. That is the convention of this chapter: Ancestral forms are usually identified as such; in addition, if forms are described in the *present tense* ("primates show . . ."), then existing forms are being discussed; if forms are described in the *past tense* ("nemertines enlarged their frontal organs . . ."), then primitive ancestral forms are being described.

From Primate to Fish

The fossil record, together with the comparative anatomy and physiology of existing organisms, has made our vertebrate ancestry relatively clear. The generally accepted line of ancestral forms, going backward, includes ape, monkey, lemur, insectivore, reptile, amphibian, a lobe-finned fish, primitive jawed fish, and petromyzont (lamprey, a jawless fish) (Romer, 1959).

In this sequence of vertebrate forms, enormous changes in motor structures have occurred with the development of manual dexterity, bipedal locomotion, vocal apparatus, lips, tongue, grinding jaws, limbs, and lungs. In this sequence from primate to fish, the basic structure of the brain and sense organs, in contrast, has changed relatively little. The major sensory and neural structures of animals in this line differ only moderately, but very little in basic structure or function. To be sure, stereopsis has been developed in primates with the splitting of the optic nerves and the combining of hemifields of vision in the cerebral cortex, the neocortex has enlarged in the mammals, the labyrinth has become an auditory as well as a body-position-and-movement detector in the amphibians and reptiles, the lateral line has disappeared in reptiles and mammals, the pineal eye has disappeared in mammals, etc. Nevertheless, the sense organs and basic neural organization found in ourselves are found in the whole series from man back to the primitive fishes; the knowledge apparatus, therefore, is very similar from us to fish (Herrick, 1948; Sarnat & Netsky, 1981). All vertebrates have eyes, ears, olfactory organs, taste buds, and basically similar brains and tubular central nervous sytems.

From Fish to Ciliate

In contrast with the line from man to fishes, the still earlier ancestry of man's own line and the relations of the vertebrates with invertebrate phyla are less certain, more speculative, and more controversial. The earlier evolution on our ancestral line occurred during the Precambrian period, for which there is little fossil record. In the absence of fossil evidence, theorization can be guided only by the facts of comparative biology. The widely taught acraniate-chordate theory of vertebrate origins (Berrill, 1955; Gans & Northcutt, 1983) focuses on certain similarities in embryology and mesodermal development in identifying as the most closely related organism *Amphioxus*, an animal that lacks the brain and sense organs of typical vertebrates. These similarities indicate affinity, but not necessarily ancestry, and the position taken here is that acraniate ancestry is unlikely. Specifically, it seems unlikely that the distance receptors and brain, the most complicated parts of the vertebrate body that so slowly changed in the evolutionary sequence from fish to primate, would have appeared suddenly in the single step from *Amphioxus* to fish. A further complication of the acraniate-chordate theory is that it connects fish with other chordates, but it specifies no particular relationship of chordates to any invertebrate phylum and so leaves unanswered the question of still earlier ancestry.

An alternative view is that acraniate chordates are degenerate relatives of vertebrates, rather than primitive and ancestral, and lost their head, brain, and sense organs in the process of degenerate adaptation for a filter-feeding, semi-sessile life-style. This alternative view offers a different ancestral line, a set of ancestors (some vertebrate and some invertebrate) in which paced changes in sensory and neural structures occurred and in which early progressive evolution of the knowledge apparatus can be followed. This alternative evolutionary sequence (the myxinoid-hoplonemertine theory, Jensen, 1960, 1963a, 1963b, 1983, 1984, 1988) bypasses the acraniate chordates and relates petromyzonts (lampreys and their kin) to the myxinoid jawless fishes (hagfishes), and these, in turn, to a complex carnivorous invertebrate, the hoplonemertine (the most advanced order of phylum Nemertini or Rhynchocoela: Burger, 1897-1907; Hyman, 1951). This theory is clearly testable (Jensen, 1988), and the tools of molecular biology (i.e., Field et al., 1988; Sibley & Ahlquist, 1983; Wilson, Carlson, & White, 1977) may soon be brought to bear upon the problem of vertebrate origins. Meanwhile, it seems reasonable to consider the implications of the myxinoid-hoplonemertine theory of vertebrate origins for the evolution of the vertebrate knowledge apparatus.

The ancestral organisms to be considered are all more primitive than the jawed fishes and are:

Petromyzonts (lampreys and ammocoete larva)
Myxinoids (hagfishes)
Hoplonemertines
Heteronemertines
Paleonemertines
Rhabdocoelous flatworms
Acoelous flatworms
Multinucleate ciliate protozoa

Petromyzonts. Petromyzonts and myxinoids are present and prominent very early in the fossil record, but these groups are represented by only a few living forms. These two groups are normally placed in the same vertebrate order (Cyclostomata) but they are very different animals indeed. The petromyzont or lamprey is the more familiar form. It lives in fresh water lakes as well as oceans and is characterized by a round, sucker-like mouth with which it can attach to rocks or to other fish, on which it feeds by rasping out flesh and ingesting blood and body fluids of its prey, typically teleost fish. It has no pectoral or pelvic fins and swims by sinusoidal undulations of its long, nearly tubular body. Its sensory and neural apparatus is very similar to that of higher vertebrates, although its labyrinth has only two, rather than three, semicircular canals. The petromyzonts have a complex life-cycle, spending some time as a filter-feeding larva (ammocoete) before metamorphosing into the adult form with a suctorial mouth. The biochemistry and physiology of the lamprey is usually similar to that of higher vertebrates.

Myxinoids. The myxinoids, like the petromyzonts, have no limbs or jaws, but there are major differences between these two forms. The myxinoids, or hagfishes, are generally recognized as the most primitive of all living craniate vertebrates (Brodal & Fange, 1963; Foreman, Gorbman, Dodd, & Olssen, 1985; Hardisty, 1979). Although they possess a notochord, tubular brain and spinal cord, ventral and dorsal spinal nerves, closed circulatory system with a ventral heart, and other typical vertebrate characteristics, they differ from petromyzonts in so many and such profound ways that they are probably placed in the same order only for taxonomic convenience. Bashford Dean, an early American ichthyologist, suggested that the evolutionary distance between myxinoid and petromyzont was probably as great as the distance from petromyzont to man. Recently gathered evidence on the physiology and biochemistry of myxinoids confirms the exceptional primitiveness of this group (Hardisty, 1979, p. 339).

The myxinoids have paired, lateral, inverse retina eyes, but have no eye muscles nor any embyological vestige suggesting they ever had them. They have a closed circulatory system with hemoglobin and a ventral heart, but neither that nor any of the several other hearts are neurally controlled, and there is no ionic regulation between the circulatory system and the water passing through the gills. Their nervous system, although similar in basic plan to that of other fishes, lacks myelin, and innervates somatic muscles by ventral roots that are anatomically different from those in any other vertebrate. Instead of jaws or a suctorial mouth, myxinoids have a pair of raspers—toothed structures at the sides of the mouth—which act to grasp and pull prey into the foregut. The myxinoid labyrinth is a single thick torus or donut that appears to sense body position and motion by rather different mechanisms (i.e., active movement of fluids by ciliary beat as well as passive movement by body rotation) than the labyrinths of higher vertebrates.

Myxinoids lack the immune system that is found in lampreys and all higher vertebrates; they have defensive behaviors (secretion of mucus from specialized slime glands, escape from grasp by knotting and passing the knot the length of the body) unknown in other vertebrates. Their primitiveness and unattractive life-style (burrowing in mud, feeding on live annelid worms or scavenging on dead or injured fish, reacting to stress by secretion of large amounts of mucus) have probably inhibited wide study of this form, but they appear to be the most primitive vertebrate with a brain and anterior sense organs.

Hoplonemertines. While the myxinoid is a peculiar and primitive vertebrate, the hoplonemertine (the next lower ancestral form according to the myxinoid-hoplonemertine theory) is a peculiar and advanced invertebrate. In the embryology of the hoplonemertine, the blastopore is associated neither with the mouth (as in typical invertebrates) nor with the anus (as in typical vertebrates and chordates) but is separate and distinct from both mouth and anus. In the hoplonemertine the regulation of embryogenesis is also not typically invertebrate in character (i.e., spiral and determinate) but more resembles the vertebrate pattern. As in vertebrates, the hoplonemertine mouth is formed by a lateral expansion of the foregut that breaks through to produced a laterally directed split under the brain; this mouth contrasts sharply in shape and position with the mouths of other nemertines, which are anteriorly–posteriorly directed slits located well behind the brain. The hoplonemertine has a complete digestive tract (mouth, foregut, midgut, hindgut, and anus), and a closed circulatory system that, although simpler than vertebrate circulatory systems, is similar in many features (position of major vessels, presence of hemoglobin, etc.). Hoplonemertines possess a complicated mesodermal system (the proboscis apparatus) that appears to have provided the rudiments for the vertebrate notochord, anterior somites, and pituitary gland (Jensen, 1963a, 1963b).

Like myxinoids, hoplonemertines are often active predators and scavengers. Some species (pelagic hoplonemertines) swim by vertical undulations of the stiffened body; although the direction of undulations is different in nemertines and vertebrates, the presence of locomotion by muscular contractions in both is noteworthy. Most nemertines, and all more primitive forms in the ancestral line being suggested, locomote by the beat of cilia covering the entire body; the use of muscles for locomotion occurs in many phyla (annelids, molluscs, arthropods, etc.), but only in a few groups (vertebrates, nemertines, leeches, and chaetognaths) are undulations of the stiffened body the propulsive mechanism.

In hoplonemertines the capture of prey is accomplished by a complex proboscis apparatus that makes it possible for a hoplonemertine to attack prey at a distance. The proboscis, a tubular structure normally introverted into the body and sometimes nearly as long as the body of the hoplonemertine, is everted or turned inside out by hydraulic pressure produced by contraction of a tubular proboscis sheath located above the digestive tract and transmitted through a fluid that fills the space between the proboscis sheath and the eversible proboscis. The everted proboscis is used to attack and subdue prey by means of a tooth-like stylet at the end of the eversible portion, and by poisonous secretions that are stored in an uneverted portion of the proboscis. In hoplonemertines the proboscis is pulled back inside the proboscis sheath by contraction of a special muscle connecting the venom reservoir and the posterior part of the proboscis sheath. By use of the proboscis, hoplonemertines successfully prey on organisms that are almost as large as themselves.

Hoplonemertines possess a large number of different specialized sense organs, many of which appear to be homologous rudiments of the vertebrate sense organs. The hoplonemertine lateral eyes, medial eyes, frontal organ, cerebral organs, head grooves, and onion-shaped organs can be considered to be homologous, respectively, to the vertebrate lateral inverse-retina eyes, medial direct-retina eyes, olfactory organ, labyrinths, lateral line, and taste buds. The hoplonemertine head is packed with neural tissue as well as multiple sense organs. The nervous tissue is organized into a solid brain that lies inside the body wall musculature and that resembles the vertebrate brain in functional organization (dorsal and ventral tissues being sensory and motor in function, respectively). The vertebrate hollow brain and spinal cord appear to have developed by the enlargement and caudal extension of a hoplonemertine sense organ, fusion with the hoplonemertine solid brain, and the extension of the fused structure as the hollow brain and spinal cord (Jensen, 1983). The hoplonemertine sense organ referred to (the chemoreceptive frontal organ) has an embryology similar to that of the vertebrate neural plate (both structures invaginate as a groove and close over to form a tube), and the hoplonemertine brain has an embryology similar to that of the vertebrate neural crest (both structures develop from special head placodes). The processes assumed by this derivation of the vertebrate brain (fusion of material from a tubular organ with material from solid placodes) are, of course, precisely what occurs in embryogenesis of the vertebrate hollow central nervous system.

Earlier Ancestors. Hoplonemertines appear to have evolved from heteronemertines and paleonemertines, more primitive orders of the phylum Nemertini or Rhynchocoela. These orders differ from hoplonemertines by having a more primitive, posteriorly situated mouth, a simpler proboscis apparatus without stylet, reservoir, or retractor muscle, smaller sense organs, and nervous systems that are located inside the muscle layers of the body wall (in heteronemertines) or subepidermally (in paleonemertines) (Hyman, 1951).

Rhabdocoels are flatworms from which nemertines appear to have evolved (Hyman, 1951); they are characterized by simpler prey capture devices than are found in nemertines, by still simpler and smaller sense organs, by a variety of simple subepidermal nervous systems, and by the absence of an anus or complete digestive tract. Locomotion is by ciliation distributed over the entire external body surface.

Acoels are the most primitive flatworms from which all other flatworms are probably evolved. Acoels have very few sense organs, minimal nervous tissue, and a syncytium rather than a cellularized gut or digestive tract. Locomotion is by cilia, but they show behavioral bilaterality, turning the anterior portion of the body left and right and investigating their surroundings by searching movements of the anterior portion of the body.

Ciliate protozoa are still simpler in morphology, being acellular in construction. They locomote by ciliary action, as do acoels and rhabdocoels, but they lack any capacity for bilaterally differentiated behavior. They turn in only one direction (typically aborally or away from the mouth) by ciliary reversal and contract their bodies in only one direction (typically toward the mouth or gullet) (Jensen, 1959).

Epistemological Significance

We have now followed the human ancestral line backward from jawless fish to ciliate protozoan. What can be said about this sequence of possible ancestors to clarify the topic of evolutionary epistemology? The first clear trend is a steady increase in size, from millimeter-long protozoa to centimeter-long rhabdocoels to decimeter-long nemertines to meter-long vertebrates. This increase in size was paralleled by increases in predatory capacity and efficiency.

When the sequence as a whole is considered, it is also clear that the early evolution of sensory organs and integrative nervous systems cannot usefully be separated from the evolution of motoric capacities. Sense organs, brains, and effectors coevolve and set the conditions and context for their mutual development. Sense organs increased in size and number in the nemertines, evolving to better control the eversion of the nemertine proboscis. The slowly locomoting flatworm can have little use for information about distant prey that it cannot chase down, contact, or capture, but the nemertine, with an eversible proboscis that can entangle and subdue prey some distance away, can use the information from eyes, chemoreceptors, and water movement detectors to control the firing of its one-shot rifle—the eversible proboscis that can be explosively thrown out to snare prey, but which can only slowly be retrieved and reintroverted. The sense organs that we normally think of as human may be of very ancient lineage, being found in the entire vertebrate line and in the nemertines as well. Our human sense organs have been refined enormously during their evolution, but their origin is probably best understood as an adaptation demanded by a change in motoric capacity, by the development of predation at a distance in the nemertines.

It is also clear that the development of primitive vertebrate neural organization (hollow brain and spinal cord extending back from the head) is probably an adaptation for lateral sinusoidal undulatory swimming. Although the brain coordinates information gathered by the many different sense organs of the vertebrate head, the main business of the vertebrate brain stem and spinal cord is the coordination of the waves of action and inhibition that sweep up and down the many body muscles, creating in the primitive fishes an efficient swimming motion. Similarly, certain developments of sensory structures within the lower vertebrates are related to the vertebrate pattern of locomotion. Modifications of the labyrinth and the development of eye muscles in petromyzonts appear to be adaptations to undulatory swimming. Complex eyes adapted for image formation can be useful in, and are found in, some slowly locomoting hoplonemertines, but such eyes are unlikely to be functional in the myxinoids, because, without eye muscles with which to compensate for head movement, lateral undulations would change and degrade the visual input from image-producing eyes and provide visual information as useful as the movies taken by a wildly waving camera. It is only after eye muscles and labyrinthine reflexes controlling the eye muscles are developed, only after it is possible to compensate for locomotor activity and to provide a stable visual field, that eyes can become effective sensory channels for guidance of vertebrate locomotion and predation.

Finally, predation is of overriding importance in understanding the evolution of the vertebrate knowledge apparatus. The sense organs, neural machinery, and effector organs that developed in the sequence leading from ciliate protozoan to primate were shaped to enhance our ancestors' potential for predation. The idea that man evolved as a predator and scavenger has been applied to recent evolution of human from ape (Dart, 1959), but it seems to describe our earlier ancestry as well. It may well be that the entire line of ancestors, from primate to fish to protozoan, is a line of predators and scavengers, and that although living representatives of ancestral groups are often rather specialized predators, the ancestors themselves, *when they first evolved*, were the generalized king-predators of their time—the largest, most motile, most capable predators and scavengers in existence. According to this view, we are the terminal twig of a branch of evolution that includes only generalized predators and scavengers and that steadily increased in size, motoric ability, sensory capacities, and neural complexity. It is the ecological niche of scavenger and predator that has created the vertebrate knowledge apparatus.

SUMMARY

Our knowledge apparatus evolved from invertebrate sense organs and neural structures that existed in nemertine ancestors. Sensory and neural structures coevolved with motor capacities, and the general trend in the evolution on our ancestral line has been the improvement of predatory abilities by the coevolutionary development of sensory, neural, and motoric structures. Distance receptors developed only as predatory attack from a distance became feasible and effective vision became possible in fish only with the development of compensatory labyrinthine reflexes that stabilize the visual field during sinusoidal swimming. The knowledge apparatus must be conceived of as an accumulation of adaptations for effective sensing of potential prey in a long evolutionary sequence of active predators and scavengers that have increased in size, complexity, and behavioral efficiency.

REFERENCES

Berrill, N. J. (1955). *The origin of vertebrates*. Oxford: Clarendon.

Brodal, A., & Fange, R. (1963). *The biology of myxine*. Oslo: Universitetsforlaget.

Burger, O. (1897–1907). Nemertini (Schnurwurmer). *Bronn's Klassen, 4 (Suppl.:I–VII), 1–542).*

Campbell, D. T. (1960). Blind variation and selective retention in creative thought as in other knowledge processes. *Psychological Review*, *67*, 380–400.

Campbell, D. T. (1974). Evolutionary epistomology. In P. A. Schilpp (Ed.), *The philosophy of Karl Popper, 14, I & II of The library of living philosophers* (pp. 413–463). La Salle, IL: Open Court.

Dart, R. (1959). *Adventures with the missing link*. New York: Viking.

Field, K. G., Olsen, G. J., Lane, D. J., Giovannoni, S. J., Ghiselin, M. T., Raff, E. C., Pace, N. R., & Raff, R. A. (1988). Molecular phylogeny of the animal kingdom. *Science*, *239*, 748–753.

Foreman, R. E., Gorbman, A., Dodd, R. M., & Olssen, R. (Eds) (1985). *Evolutionary biology of primitive fishes*. New York: Plenum Press.

Gans, C., & Northcutt, R. G. (1983). Neural crest and the origin of vertebrates: A new head. *Science*, *220*, 268–274.

Hanson, E. D. (1977). *The origin and early evolution of animals*. Middletown, CT: Wesleyan University Press.

Hardisty, M. W. (1979). *Biology of the cyclostomes*. London: Chapman & Hall.

Herrick, C. J. (1948). *The brain of the tiger salamander*. Chicago: University of Chicago Press.

Hyman, L. (1951). *The invertebrates: Vol. II. Platyhelminthes and Rhynchocoela—The acoelomate bilateria*. New York: McGraw-Hill.

Jensen, D. D. (1959). A theory of the behavior of *Paramecium aurelia* and behavioral effects of feeding, fission, and ultraviolet microbeam irradiation. *Behaviour, IV*, 82–122.

Jensen, D. D. (1960). Hoplonemertines, myxinoids, and deuterostome origins. *Nature, 187*, 649–650.

Jensen, D. D. (1963a). Hoplonemertines, myxinoids, and vertebrate origins. In E. C. Doughterty, Z. N. Brown, E. D. Hanson, & W. D. Hartman (Eds.), *The lower metazoa: Comparative biology and phylogeny*. Berkeley: University of California Press.

Jensen, D. D. (1963b). Hoplonemertine ancestry and vertebrate head structure. *Proceedings of the XVI International Congress of Zoology*, Washington, DC, 192.

Jensen, D. D. (1983). Hoplonemertean ancestry of vertebrates: An old head and new brain for craniates. *American Zoologist, 23*, Abstract No. 284.

Jensen, D. D. (1984). Predation and the origin of vertebrates: A comparison of recent theories. *American Zoologist, 24*, Abstract No. 452.

Jensen, D. D. (1987, August). *Operationism, polytheticism, the biotemporal space, and the empirical integration of ethology, psychology, neuroscience, and sociobiology*. Poster, International Ethological Conference, Madison, WI.

Jensen, D. D. (1988). Hubrecht, Macfarlane, Jensen, and Willmer: On the nature and testability of four versions of the nemertean theory of vertebrate origins. *Hydrobiologia, 156*, 99–104.

Medicus, G. (1987). Toward an etho-psychology: A phylogenetic tree of behavioral capabilities proposed as a common basis for communication between current theories in psychology and psychiatry. *Ethology and Sociobiology, 8*, 131S–150S.

Romer, A. S. (1959). *The vertebrate story*. Chicago: University of Chicago Press.

Sarnat, H. B., & Netsky, M. G. (1981). *Evolution of the nervous system* (2nd ed.). New York: Oxford University Press.

Schneirla, T. C. (1949). Levels in the psychological capacities of animals. In R. W. Sellars, V. J. McGill, & M. Farber (Eds.), *Philosophy in the future* (pp. 243–286). New York: MacMillan.

Sibley, C. G., & Ahlquist, J. E. (1983). Phylogeny and classification of birds based on the data of DNA-DNA hybridization. In R. F. Johnston (Ed.), *Current ornithology, I* (pp. 245–292). New York: Plenum Press.

Thomas, R. K. (1986). Vertebrate intelligence: A review of the laboratory research. In R. J. Hoage & L. Goldman, *Animal intelligence: Insights into the animal mind* (pp. 37–56). Washington, DC: Smithsonian Institution Press.

Wilson, A. C., Carlson, S. S., & White, T. J. (1977). Biochemical evolution. *Annual Review of Biochemistry, 46*, 573–639.

9 Evolutionary Epistemology Naturalized: Comments on Lerner, Danailov, and Togel

Robert Feleppa
Wichita State University

In keeping with my professional background in philosophy, I do not comment on matters of biological detail here. Instead, I call for clarifications of argument on several points in these chapters, while confining my more constructive comments to general epistemological and metaphysical aspects of these chapters. Specifically, I make an effort in a number of places to call attention to the relevance of contemporary philosophical pragmatism, particularly the work of W.V. Quine, to these discussions.

Turning first to Lerner, I have trouble seeing exactly why the probabilistic epigenesis or contextual-developmental account is any less likely to be used to defend the idea of privileged cognitive systems than the deterministic epigenesis account. The middle section of Professor Lerner's chapter demonstrates convincingly (to me, anyway) that there are serious problems in trying to account for the development of all new features in terms of hereditarily fixed processes that burden the explainer with the task of showing the evident survival value of each of nature's "unmistaken" developmental revisions. However, what is to prevent one from adding to the strictly adaptive features to be enhanced by harsh social policy a catalog of exaptive ones? Is it just that they are not as readily grasped as traditionally adaptive ones? I would assume that Lerner's ideal theory of cognitive development would offer probabilistic laws concerning advantageous patterns of environmental and social interaction for cognitive development. As the latter are not nearly as well understood as the more traditional processes of biological adaptation, perhaps eugenic policies become less of a *present* threat, however. But in advocating a "more plastic" evolutionary epistemology, based on probabilistic rather than deterministic laws, we should not allow ourselves to connect this necessarily with the idea that there are strong inherent limitations to the predictive power of developmental-contextual components of future theories. And the albeit important ethical matters at issue here should not induce us to make such an inference either.

Aside from this concern, I am largely supportive of Lerner's position, and let me offer some—I hope helpful—supplementary considerations. For one, biological reductions of culture have to this point left just too many outstanding promissory notes to make a convincing case for downplaying the causal role of human-symbolic interaction in the development of culture and cognition. Also, some existing programs for evolutionary reduction of culture and cognition seem to require linguistic and psycholinguistic theories that attribute a lot of genetic hardwiring of linguistic categories to the brain; that is, such determinism brings extensive assumptions about shared cognitive structure to bear on linguistic analysis—if it is to succeed in severe biological reduction of culture—and it is not clear that this is the most viable of linguistic strategies. Quine prefers, as do I, a linguistic theory that makes only minimal assumptions about underlying physiological or mentalistic uniformity in accounting for the behavioral uniformity of language. Quine attributes a lot to social-behavioral conditioning and as such invokes considerations potentially antagonistic to strong reductionist programs of the type Lerner criticizes. As Quine aptly summarizes it at one point, linguistic uniformity is achieved in ways more analogous to the trimming of hedges to similar shapes. External uniformity is achieved without corresponding internal uniformity (of twig structure.) Although Quine's naturalistic philosophy welcomes the information of philosophical questions by neurophysiology, evolutionary biology, and other sciences, and although he is specifically concerned to put human science and cognitive development in evolutionary perspective, much of his work (like Wittgenstein's) aims at emphasizing the social context of language learning and its antipathy to mentalistic and rationalistic theories of meaning and language acquisition (Quine 1960, 1973, 1975, 1980).

The compatibility of Quine's position with Lerner's lies in its downplaying the role of inherited physiology in language learning whereas emphasizing the individual's interaction with the natural and social environment. This comes particularly close to the surface in Lerner's discussion of Annis, insofar as Annis's notion of the "objective pull of nature" is derived from Quine (see Annis, 1982). However, Lerner is critical of Annis for reasons that are unclear to me. Annis does indeed give a different emphasis to the idea of objective pull than Quine, insofar as he specifies a pull "of nature" whereas Quine's initial discussion indicates a largely *social* one. Quine's idea is that our conditioning to apply words to various things must deviate from simple association of words with similar repeated stimulations, and that this is achieved mainly by the help of our human teachers. For instance, although the learning of a color expression can be based largely on being conditioned to respond to similar stimuli, our learning of "physical shape" cannot: We must learn to call "square" what might well appear to us, because of our varying perspectives, as having various trapezoidal forms. This latter is what Quine calls *objective pull*—away from simple subjective association to properties acknowledged as attributable to the "objects themselves," because they cannot be attributed to items in the sensory field. Quine maintains that social conditioning has much impact on this attitude and on our ontological and epistemological categories in general.

Annis's discussion emphasizes the impact of environmental factors on this conditioning process. For instance, there is evident survival value in Eskimos being able to make exceptionally fine visual discriminations in their largely uniform perceptual environment, and Annis (1982) argues that this particularly well illustrates the objective pull of nature:

> In order to survive in their harsh environment, the Eskimo had to develop acute perceptual skills that others were not required to develop. The development of these skills in part may be explained by natural selection. Those subjects that displayed more discriminatory behavior survived and hence these skills were genetically reinforced. When we view the development of the human species over time, it is clear that evolution has played an important part in the development of our epistemic practices . . . , one external constraint on an epistemic practice is its survival value. (p. 205)

Lerner sees in this something antipathetic to his concerns, but I am not sure why. Annis' emphasis is not sharply biological but seems perfectly continuous with the blend of social, environmental, and biological that we find in Quine and Lerner. Far from being deterministic, Annis (1982) argues that "social and cultural features . . . are not determined by the ecology but the general pattern of cultural adaptation is probabilistically related to the ecology" (p. 205). At any rate, I don't think Quine would object to Annis' adaptation of "objective pull."

As for Professors Danailov and Togel's chapter, my concerns center on the exact nature and scientific import of the various notions of realism and correspondence discussed by them and in related literature.

First, they and others such as Campbell, Lorenz, and Vollmer wish to emphasize the "mind independence" of the world, whereas also emphasizing a certain dependence of mind on the world—or, to use a locution of Vollmer's employed in both chapters in this volume, the "partial isomorphism" of subjective cognitive structures to the world. I think I agree that these are desiderata, but I wonder at the nature of the philosophical commitments involved.

Mind independence is tied to realism and to a correspondence theory of truth. Typically presented as a wrong-headed view of things is idealism, insofar as it makes reality too mind dependent and tends to favor an account of truth in terms of the coherence of beliefs with each other as guided by basic epistemic standards. Now I would agree that there is something wrong with idealist theses of mind dependence that might deny that the world can exist without minds of some kind. But idealists are also wont (in fact more wont) to stress mind dependence in another important sense—roughly, that human cognitive categories are partially constitutive of reality: or, to put it more bluntly, that mind does not "mirror" nature but imposes structure on it.

I think this thesis is much stronger, though I prefer the version evolved in the post-Kantian tradition of pragmatism, which stresses theory's role as an instrument for the attainment of various human aims. Much as the idealists talk about cognitive categories being imposed on "reality," pragmatists talk of human need and purpose being imposed. More specifically, pragmatists have much difficulty with the idea that the categorial terms of theory can be proven, from some supratheoretical, a priori perspective, to match the categories of nature, its so-called "natural kinds." Convention and historical contingency play too strong a role in the development of theory for this to hold up.

This is not to say that world-theory correspondence cannot be described or theoretically understood. For a number of pragmatists, and notably Quine, scientific theory is allowed to tell us what it can about the world and about the relations of thought and theory to the world. Here Quine follows largely in the footsteps of C.S. Peirce in viewing ultimate standards of truth as to be articulated *within* scientific theory. Quine's account of criteria for evaluating proposed natural kind schemata explicitly cites survival value as a measuring rod. (But, not to create apparent conflict with earlier remarks, it provides only the most general of constraints on natural kind terms: A full wiring of linguistic ability is not intended.) And this account is much in line with a general thesis of "internal realism" (i.e., that all talk of what is real or what is a fact—physics and metaphysics in the fullest sense—is done within what Quine terms the *ultimate parameters* of scientific theory). There just is no higher standard to appeal to in telling how the world is, for Quine, than what scientific theory tells us.

As there are many theories of what the world is like, and as it is not clear how they can be regarded as equivalent to each other, most internal realists hold that there is more than one true description of the world. The contrasting metaphysical realism, on the other hand, projects a world consisting, as Hilary Putnam (1981) remarks, "of some fixed totality of mind-independent objects. There is exactly one true and complete description of 'the way the world is.' Truth involves some sort of correspondence relation between words or thought-signs and external things and sets of things." it adopts, in effect, "a God's Eye point of view" (p. 49). Campbell's "hypothetical realism" might seem to have a foot in both metaphysical and internal realist camps. It is largely internal, but it emphasizes a mind-independent reality. But as this is a reality whose nature and existence are determined by scientific argument, it is not metaphysical or "external" in the sense I mean here. The key point of departure for metaphysical realism lies precisely in its resting its realism on extrascientific considerations.

Quine specifically assigns to *physical* theory a special ontological status (i.e., that theoretical differences or disagreements in any scientific theory are *factual* ones only if there is a corresponding difference in physical state). However, other pragmatists, notably Quine's Harvard colleague Nelson Goodman, are not as wont to place physics in such a privileged position (Quine, 1981). Goodman advocates more a Jamesian, pluralist pragmatism that holds nonscientific systems to be of equal merit to the scientific—"equal" because the differing needs they meet yield differing criteria of adequacy with little or no cross-system validity. Goodman particularly stresses the equality of artistic and scientific systems (or *versions* as he calls them) and argues that there are as many worlds as there are adequate versions (see e.g., Goodman 1978, esp. Chaps. 1, 6, 7). Also, in contrast to Quine, he eschews even a naturalistic defense of the notion of natural kinds, defending instead a radically conventionalist position regarding the criteria of adequacy of scientific categories (Goodman, 1978, Chap. 7; 1973, esp. Chaps. 3–4).

However, more important for the present context are the similarities between Quine and Goodman and not their differences—particularly as regards their rejection of metaphysical realism. And what they share may present a challenge to the accounting of *Gegenstandlichkeit* that Professors Danailov and Togel see as missing from internal and hypothetical realisms, and as reason to favor what appears to be a metaphysical realism. The key idea is that many traditional epistemological and metaphysical positions seem to hinge a lot on the idea that we can somehow make a warranted comparison between our cognitive descriptive products and the "world itself." But Kant, Hegel, Peirce, James, Cassirer, the Gestalt psychologists, and more recently Wittgenstein and Kuhn, among others have raised a host of telling challenges to this idea. The germ of the problem, I think, is contained in George Berkeley's objections to modern-era critical (and naive) realisms that assert or deny similarity between properties of sense-impressions and external properties; that is, how are we ever in a position to make the requisite comparison between what our senses tell us and what the world is actually like—because our senses seem to give us all the information we can have? More generally, how do we compare our various accounts of the world with "the world itself," if the latter is construed as being apprehended without the benefit of cognition? Can we make sense of this notion, and if we can, what epistemological work can it do for us? We can say, of course, that the world "is there" and that it exists independently of our efforts to conceptualize it—but if we try to make more point to point comparisons, to spell out the "partial isomorphism" of conceptual scheme and world, we seem necessarily to be comparing descriptions. The more we consider the underlying reality that our various schemes describe or organize, the more it appears "evanescent." Goodman (1978) queried: "But what is *it* that is so organized? When we strip off as layers of convention all differences among ways of describing *it*, what is left? The onion is peeled down to its empty core" (p. 118).

Although Quine and Goodman clearly differ on the matter of the priority of physical theory, Goodman's arguments lend weight to Quine's rejection of skepticism and philosophical a priorism. Campbell (1987) remarks that naturalistic epistemologists such as Quine, and evolutionary epistemologists (and hypothetical realists) such as he himself, "deliberately beg the question" raised by traditional epistemologists "and undertake epistemological inquiries assuming that present-day physics and astronomy give us approximately valid knowledge of the world to be known, and that evolutionary biology, psychology, and sociology tell us something about man the knower" (p. 165). However, Goodman raises serious challenge to the idea that there is any question to beg.

Of course various strategies have been employed to give justification to the idea that true theory mirrors nature. Some of the figures I just mentioned attack, I think successfully, the most plausible way of doing this, namely by asserting that a correspondence of the observational claims of theory to the world can be warranted without appeal to current theory. Others follow more Kant's line of exhausting various rationalist strategies for showing world-theory correspondence and find much the same dissatisfaction that Professors Danailov and Togel show with the idea that the world is independently there, but that's about all we can say: particularly if this is seen as entailing any kind of restriction at all about what can be said about the connection between cognition and reality.

If we agree with them that the world-in-itself must be cognitively accessible, and thus that the inaccessible Kantian world is best passed over in silence, what else is there to say beyond what internal realists allow us to say about theory–world relations? What about the Gegenstandlichkeit of theory and world do they neglect?

The pragmatist verdict is this: If one wants a world to "stand against" our theory, what is wrong with the theory's account of that world. The language used to describe the world can also function as a metalanguage describing the relation of the theory to the world. Whether it succeeds in this task is to be determined according to the criteria of adequacy that competent practitioners in the field apply to anything they do. These criteria can themselves be partly evaluated in terms of other theories—say of rational choice. But if one asks for a final justification of all the criteria, one is asking, in effect, for a justification that appeals to no criteria at all—or ones whose validity is just rationally certain. But these avenues do not seem fruitful. We work within our conceptual scheme and really cannot get outside it to tell if it is really working. The mere logical possibility that we might after all be entirely wrong is not a basis for serious philosophical doubt, much less serious scientific doubt.

One cannot have an a priori proof that science is trustworthy. And for this reason, Quine argues, epistemology must be "naturalized" in that it must be wrested from the clutches of a priori philosophy and viewed as a component of science. And the same goes for metaphysical theses about how the world in fact is, that is, implicit here is an abandonment of First Philosophy and its efforts at finding a priori, rationally indubitable starting points for the justification of all belief.

Quine (1973), as I said, offers an account of natural kinds, and of the more general concerns of Hume about induction, that appeals to natural selection. The following passage ties this evolutionary epistemology to a naturalized epistemology: "Asking for a justification of induction is like asking for a first philosophy in support of science. What natural selection contributes, rather, . . . is a reason why induction works, granted that it does. We have here, indeed, an illustrative part of an answer to . . . the central question of enlightened epistemology: how, if our theory of the external world is true, could we ever hit upon it?" (p. 20).

I do not deny that it is of obvious importance for accounts of cognitive development and adaptability to relate cognition to the world. But need one thus be committed to problematic aspects of correspondence realism? If the criteria of adequacy of evolutionary biology, or of science generally, are met, why is this not sufficient? The pragmatist account provides the most fruitful avenue of escape from these difficulties, whereas at the same time embodying the insights of Engels, Weizsacker, and Simpson at its very heart: Theories are ultimately measured not by mirroring reality but by meeting the useful purposes to which they are put.

REFERENCES

Annis, D. B. (1982). Epistemology naturalized. *Metaphilosophy, 13*, 201–208.

Campbell, D. T. (1987). Neurological embodiments of belief and the gaps in the fit of phenomena to noumena. In A. Shimony & D. Nails (Eds.), *Naturalistic epistemology* (pp. 165–192). Dordrecht, Holland: D. Reidel.

Goodman, N. (1973). *Fact, fiction, and forecast* (3rd ed.). Indianapolis: Bobbs-Merrill.

Goodman, N. (1978). *Ways of worldmaking*. Indianapolis: Hackett.

Putnam, H. (1981). *Reason, truth, and history*. Cambridge: Cambridge University Press.

Quine, W. van O. (1960). *Word and object*. Cambridge, MA: MIT Press.

Quine, W. van O. (1973). *The roots of reference*. LaSalle: Open Court.

Quine, W. van O. (1975). The nature of natural knowledge. In S. Guttenplan (Ed.), *Mind and language* (pp. 67–81). Oxford: Oxford University Press.

Quine, W. van O. (1980). Sellars on behaviorism language and meaning. *Pacific Philosophical Quarterly, 61*, 26–30.

Quine, W. van O. (1981) Goodman's ways of worldmaking. In W. Quine (Ed.), *Theories and things* (pp. 96–99). Cambridge: MA: Harvard University Press.

10 The Amoeba and Einstein

Ethel Tobach
American Musem of Natural History
New York

The participants of this volume discuss the evolution of knowing by examining the approaches of evolutionary epistemology and integrative levels. According to evolutionary epistemology, the ultimate cause of differences between the "knowing" of an amoeba and an Einstein is the result of the inheritance of different structures evolved by the species, through a selection process.

Schneirla was opposed to vitalistic explanations as expressed in the concept of inherited behavior patterns, in anthropomorphism, and in nondevelopmental approaches to behavior (Aronson, Tobach, Rosenblatt, & Lehrman, 1972). These concerns are exemplified on pages 445–446 and 479 in the classic book he coauthored with N.R.F. Maier (Maier & Schneirla, 1935). The organization of the book, and its stress on valid and reliable research to experimentally define the similarities and differences between the higher mental processes in humans and those in animals other than humans, demonstrate an early use of the levels concept.

Campbell (this volume) constructively and thoughtfully relates evolutionary epistemology to Schneirla's "several levels of integration themes." The depth of Campbell's thinking and the breadth of his scholarship contributed much to the possibility of achieving the aim of this book and challenged us to heed the call for "more refinement and reconceptualization" with which he ends his chapter.

The further development of Schneirla's concept of integrative levels was unfortunately aborted by his untimely death. Since that time, others (Aronson, Greenberg, Leacock, Tobach, & Topoff) have presented their own views of the concept of integrative levels (Greenberg & Tobach, 1984, 1987, 1988) Although all acknowledge their debt to Needham (1929), Novikoff (1945), and Schneirla (1949), these five do not agree in all respects with each other. So, the choice of a statement of "Schneirla's concept of integrative levels" involves not only a consideration of his own development, but of the contributions of the others. The paper by Schneirla on approach–withdrawal processes (1965) is most pertinent to this discussion. To focus the discussion further, I refer primarily to Campbell's chapter in this volume. I have presented many of the following formulations previously (Tobach, 1981, 1987), but they need restatement in the context of this book.

The frequent use of *hierarchies* and *levels* interchangeably leads to a need for clarification of the terms. The commonsense definition of *hierarchy*, that is, of arrangement by rank, takes on a clear connotation of structure determined by some external agency, such as the church, army, government, or other societal institutionalization of what is considered orderly. The emphasis in hierarchy is structure, and the relations between and among the ranks is indeed ordinal. The ranking in a hierarchy does not depend on the inherent characteristics of change in a lower rank to produce the higher rank. The use of nested hierarchy for knowledge processes results in serious differences between Campbell's analysis of evolutionary epistemology and the evolution of knowing or epistemics (in the sense of act or ways of knowing) derived from the concept of integrative levels.

There is no space here for a historical analysis of the use of the term *level* by Needham, Novikoff, and Schneirla. However, the concept of "level" from physics as an unstable, quantized energy state comes close to their conceptualization. Campbell's (1974) use of "semistable molecular arrangements" in his selective elimination model comes close to this view. The ordering of integrative levels depends on changes in their entities, components, or units. The instability is relative and reflects the change characteristics of the level. The quantification is persistent to the characteristics of the energy form .of the level (physical particles; cells; organisms).

The ordering is not in a ranking sense, but in a process-time sequencing sense. Levels are ordered or arranged "integratively" rather than as structural "levels of organization." Any level subsumes (an)other(s) by an integration (fusion) of the changes of the other level(s) (external and internal relationships) that produce the new level. The concept of "integration" is critical for the use of "levels" in understanding the ordering of matter.

Equally relevant is the difference between "emergence" and "production of a new level." The connotation of "emergence" is that the previous or lower level contains that which now "emerges." This is only real in part: The new level has aspects or components of the previous level(s), but it also has aspects or components that are not present in the previous level(s).

In the opening section of his chapter, Campbell declares that he shares four themes with Schneirla: (1) integrative levels; (2) materialism, avoiding atomistic reductionism, (3) avoiding vitalism and teleology; (4) reconciliation of the autonomy of each level with the fact that biological wholes determine the nature of the parts, and the parts determine the nature of the wholes. Themes 2, 3, and 4 are discussed as they are pertinent to other issues.

Campbell presents some concepts that are similar to those of integrative levels; however, when these concepts are embedded in his basic premises of downward causation, selection-theory, and nested hierarchies, the differences between Campbell's approach and integrative levels become evident. Some examples follow.

Figure 1.1 (Campbell, this volume) demonstrates the significance of a fundamental requirement of the integrative levels concept: designation of the category in which the levels are to be delineated and analyzed. One might consider Campbell's term *dimensions* to signify *category* in the sense that his term *Complex Level of Organization* could be seen to have different levels of complexity that lead to increasing complexity of organization. Or, his term *evolutionary order* could be a category, in the sense of phylogenetic systematics. Within that category, different behavioral/structural levels of organization could be delineated. From Campbell's figure, it is not clear whether all behavioral categories are considered equivalent. It is also not clear whether the structure relates to individuals or taxonomic groups.

The need to designate the category being discussed is evident in Campbell's statement that, whereas Schneirla and Topoff reject the concept of "the superorganism," they accept the reality, that is, the ant colony level of organization. Novikoff (1945) and Schneirla (1951) made it clear that it is impossible to designate a level of social organization as "superorganismic" within the category of social organization.

Campbell's "Shared Theme 3, Avoiding Idealistic Vitalism and Teleological Causation," is worthwhile discussing in the light of the second of Campbell's four principles of levels of organization concerning teleonomic achievement at higher levels derived from micromechanisms on lower levels. He views this process as a "reductionist" constraint. Teleonomy (the concept that an organism's structures or functions must have given it an evolutionary advantage, Webster's World Dictionary) rests on an adaptationist view of evolution. The tautological aspect of this view has been dealt with extensively (Ho, 1987). To avoid the problem of teleonomy, which may be considered a particular form of teleology, I would reformulate the second principle to read as follows: "explanation is not completed until (a) lower level processes have been specified within their level, and their interdependence with each other in that level were defined; (b) until the changes in their interdependence in their relations with other levels have been delineated; and (c) until these interdependencies are defined within the next higher level." It is not necessary to imply any purposiveness in the processes of evolution and development to understand them. I also believe that my reformulation obviates reductionism.

The third principle of Campbell's levels of organization, the emergentist principle (discussed earlier), states that biological evolution meanders (aimless wandering; the dictionary, 1988, gives a stream meander as an example; today there are data to support the possibility that the meander of a stream is not "aimless") while exploring parts of the universe. In these explorations, physical laws act selectively. This is a vitalistic formulation in that it reifies biological evolution. In Campbell's principle, the "emerging" animate or biological level is the result of an external process, the selection process, that obeys some higher principle that keeps the development of every level of organization in order.

The concept of integrative levels offers a materialist approach, in which the significant difference between the inanimate and animate levels lies in the qualitative leap in energy characteristics. That qualitative change from quantity to quality is inherent in the characteristics of the inanimate level (Eigen, 1971; Prigogine, 1980). This qualitative change requires definition of specific laws pertinent to that process; the explication of those laws is experimentally feasible.

Campbell's discussion of downward causation is related to his shared theme 4, "Autonomy of each level; wholes determine the nature of parts, parts determine nature of the whole." Whereas this theme is consonant with my view of the concept of integrative levels, the cardinality of downward causation as defined as the fourth principle in Campbell's levels of organization critically changes the agreement of the theme with the integrative levels concept. The causal relationship among and between levels is dialectical and bidirectional.

For example, an individual provokes an angry comment by a supervisor. In the intraorganismic category, the changes on the physiological level (endocrine and neural systems) change the function on the biochemical level (adrenal hormones and neurotransmitters), effecting a change back into the physiological level, affecting the neuromusculature, resulting in change in the facial muscles, presenting an adequate social stimulus (change) to the supervisor that plays a significant role in further development of the angry interchange. In the category of social behavior, the intraorganismic changes are integrated in the individual as a component in the level relevant to the worker–supervisor relationship. The social level subsuming all the intraorganismic levels then affects the biochemical level, as well as the physiological level. Again, the physiological level functions upward to the social level that functions downward to the physiological level and so forth.

To fully understand the processes involved on all levels, knowledge of the history of the individuals involved would also be necessary (societal level). Causal explanations are not only dialectically "upward" and "downward" but also interdependent with other categories as related to the category being investigated.

In the selection-theory and selective-retention model, the emphasis is on the elimination of genetic material, rather than on the process of mutation (change) in the configurations of the molecules. The "validation by survival" is the process of selective retention. The selective retention paradigm is applicable to a variety of physiological processes (e.g., embryological growth and wound healing) as well as to behavior in general, particularly in the four processes relating to knowledge (Campbell, 1974, p. 421). The model rests on the tautology of adaptation (or validation, as used in Campbell's model) and genetic fitness (Ho, 1987). The gene is conceptualized as the ultimate, limiting, and determining factor in evolution. The vitalistic nature of these assumptions were discussed earlier.

Although Campbell agrees with Lewontin (1982) about the centrality of the "construction" of the niche by the species, the internal selection of these constructions rests at every node on the idealist concept of the "vicarious control system . . . nested within the whole-organism-reproduction control system of natural selection" (Campbell, 1987, p. 186). This formulation by Campbell stems directly from the primacy of downward causation.

Integrating the levels of structure/function (an interdependence that Campbell recognizes) projects possible processes by which such changes come about. For example, how does the activity of the worker ant relate to changes in the jaw that are a factor in the creation of the ecological niches of the species? It is possible that ecological factors (change in the food source) and experiential factors (change in developmental patterns in relation to food changes) may bring about a change in individual biochemistry (e.g., newly introduced amino acids). With the continued change in the type of ingested material, changes in muscles and other tissues take place (the internal contradictions). The accusation of Lamarckianism is immediately heard! However, we are talking of change on the biochemical level, on the molecular configuration level. Geneticists have demonstrated many ways in which such new configurations and new functions take place (Cairns-Smith, 1986). Such microchanges are difficult to define, but it is no longer impossible to do so.

The four processes outlined in the abstract statement of the dogma underlying the nested hierarchy of knowledge processes are: (a) blind-variation and selective-retention; (b) variation based on consistent selection processes and mechanisms for propagating and preserving selected variations; (c) inductive achievements containing the wisdom achieved by blind variation and selective retention (short-cut processes); and (d) short-cut processes that substitute for overt exploration and organic evolution (Campbell, 1974, p. 421) The pathways for producing these processes are finally attributed to the "life-and-death winnowing of organic evolution."

Campbell (1974) states that: "An essential connotation of blind is that the variations emitted be independent of the environmental conditions of the occasion of their occurrence . . . the occurrence of trials individually be uncorrelated with the solution . . . A third essential connotation of blind is rejection of the notion that variation subsequent to an incorrect trial is a 'correction' of the previous trial . . ." But in the case of the behavior of an animal in a problem-solving situation, the reason that the behavior is not random is that the animal has: "1. Already achieved wisdom . . . due to inheritance and learning . . . and 2. Maladaptive restriction on the range of trials . . . (Such biases are due to, ET) structural limitations and to past habit and instinct." (p. 422).

The concept that a variation (behavioral or otherwise) can be independent of its environment reflects the centrality of heredity in Campbell's theory. The dichotomous analysis of the experience of the organism in terms of "heredity" and "environment" has been shown to be invalid (Tobach, 1972). From an integrative levels view, the experience of the individual that includes its history on all organismic levels and the adequate stimulation from the contemporaneous setting in which the events take place are always functional factors in any process. Such an approach is in contrast to that of the hereditarian view.

The primary contradiction in evolution is the activity of the organism that maintains or destroys its integrity. This implies the dialectic relationship between the continuous internal contradictions in the organism, and the external contradictions in which the organism is engaged, that is, with the abiotic and biotic processes external to the organism. The "externality" and "internality" are also continuously changing as biophysical/biochemical processes at the interface take place. There is no implication in this proposal that the status of the integrity of the organism is defined in terms of life or death, or reproduction or failure to reproduce. The major difference between this dialectic view of natural selection and the structural or vitalistic, reductionist view is that the latter relies on agencies that exist independently of these processes, agencies that direct from above the rules and possibilities of change. In the hereditarian view, variation is blind and becomes known only through the tautology of adaptation.

The revised version of nonnmemonic problem solving, the first in the nested hierarchy of knowledge processes, rests on "presumptions" about the nature of the environment in which the organism finds itself. Campbell sees the process as a continuum in the evolution of the behavior of all animal species (protozoan to primate). It is also based on a comparator function of a presumed knowledge of nourishing or noxious gradients.

In some respects, the new concept resembles the Schneirla approach/withdrawal concept (1965). There are some significant differences, however. In Campbell's view, the discrimination of the nourishing and noxious factors are based on natural selection of inherited structures. The integrative levels approach would define species-typical responses that are the integration of the total experience of the individual organism, given the fact that most organisms develop in a species typical pattern of experience. However, in the case of those organisms with time-limited storage and stimulus adequacy characteristics, the responses of the individual are based on the objective, contemporary biophysical and biochemical characteristics of the new situation in which it finds itself. Because of its activity, or internal change, its thresholds to adjust to the situation change, and the probability of increasing or decreasing activity and redirecting or not redirecting its locomotion, or other activity, also changes. The change in activity is a function of quantitative and qualitative characteristics of the external situation. The quality of nourishment or noxiousness acts through its biochemical effects on the processes underlying approach and withdrawal.

The third and fourth processes in the nested hierarchy present the concept that instinct, habit, and visual diagnosis are interlocked and interdependent. The issue of instinct has been fully discussed elsewhere (Lehrman, 1953; Schneirla, 1965).

The seventh (socially vicarious exploration) and eighth (language) processes require some comment. I have proposed an alternative approach to language elsewhere (Tobach, 1987). In the discussion of socially vicarious exploration, a typical problem ensuing from a misreading of the integrative levels concept is evidenced, that is, the results of the processes of a particular level are confused with the processes themselves.

Although it is difficult to make generalizations about "ants," a group of animals in which a variety of evolutionary patterns are evidenced, the following scenario is probable in representative species.

In some species, the chemical trail left by ants that have carried food back to the nest presents an approach stimulus that keeps the next foraging ant on the trail (Campbell's "vicarious exploration"). The chemical trail may be the result of the process of metabolic change in the food-carrying ant. The following of the trail is very different from "exploration." In exploration, the ant is responding to a variety of stimuli that may be new or that may have reference to past experience. The adequacy of the pheromone as an approach stimulus is an integration of the ant's experience on many levels and dependent on its developmental stage. The chemical trail that is put down may be different from other chemical trails because of the metabolic, neuroendocrine, and other processes on the physiological level. That chemical might not have the same effect on an ant that was returning with food to the nest, or an ant from another species, or a different colony of the same species.

The areas of agreement and disagreement between Lorenz and Campbell are well documented (Campbell, 1974, 1975). Although Campbell is critical of Wilson's sociobiology (Campbell, 1983), he has not yet addressed the culturgen concept in his writings. E. O. Wilson (1978) and Lumsden and Wilson (1981), however, indicate that Campbell's views on the relation between social and biological evolution are acceptable to them.

Evolutionary epistemology, as conceived by Campbell, and accepted by Popper (1974), and the culturgen as conceived by Lumsden and Wilson (1981), are based on the same invalid dichotomy, that is, of heredity and environment. This formulation is one that has been argued and written about extensively. Attempts have been made to deal with this dichotomy by projecting different types of "interactionism." As I have indicated elsewhere (Tobach, 1972), interaction between the two entities is a type of additivity that leads frequently to statistical studies to determine "how much is genetic and how much is environmental."

We are once again faced with the question: Why does this pattern of dichotomous thinking, which leads to an emphasis of the genetic limitations of populations, and eventually, of individuals, continue to be expressed in the scientific community? The insistence on the dominance of a genetic process that is resistant to change, and thus determinate, does not lead to any progress in our understanding of the evolution and development of behavior.

As the concept of integrative levels proposes, ideas will change, but the emphasis on the activity of organisms as the preeminent factor in that change results in a positive approach to the answer to my question. Despite all disagreements among the genetic determinists, the materialist emergentists, and the level integrationists, there is agreement that solution of these problems lies in the development of human wisdom.

REFERENCES

Aronson, L. R., Tobach, E., Rosenblat, J. S., & Lehrman, D. S. (1972). (Eds.). *Selected writings of T. C. Schneirla*. San Francisco: W. H. Freeman.

Cairns-Smith, A. G. (1986). *Seven clues to the origin of life*. Cambridge: Cambridge University Press.

Campbell, D. T. (1974). Evolutionary epistemology. In P. A. Schlipp (Ed.), *The Library of Living Philosphers*. (Vol. XIV, Book I). LaSalle, Il: Open Court.

Campbell, D. T. (1975). Reintroducing Konrad Lorenz to Psychology. In K. Lorenz; *The man and his ideas*. New York: Harcourt, Brace, Jovanovich.

Campbell, D. T. (1983). The two distinct routes beyond kin selection to ultrasociality: Implications for the humanities and social sciences. In *The nature of prosocial development: Interdisciplinary theories and strategies*. New York: Academic.

Campbell, D. T. (1987). Neurological embodiments of belief and the gap in the fit of phenomena and noumena. In A. Shimony & D. Nails (Eds.), *Naturalistic epistemology: A symposium of two decades*. (pp. 165–192). Dordrecht, Holland: D. Reidel.

Eigen, M., (1971). Selforganization of matter and the evolution of biological macromolecules. *Die Naturwissenschaften*, *58*, 465–523.

Greenberg, G. & Tobach, E. (1984). *Behavioral evolution and integrative levels*. Hillsdale, NJ: Lawrence Erlbaum Associates.

Greenberg, G., & Tobach, E. (1987). *Cognition, language, and consciousness: Integrative levels*. Hillsdale, NJ: Lawrence Erlbaum Associates.

Greenberg, G., & Tobach, E. (1988). *Evolution of social behavior and integrative levels*. Hillsdale, NJ: Lawrence Erlbaum Associates.

Ho, Mae-Wan (1987). Evolution by process, not by consequence: Implications of the new molecular genetics on development and evolution. *International Journal of Comparative Psychology*, *1*, 3–27.

Lehrman, D. S. (1953). A critique of Lorenz's "objectivistic" theory of animal behavior. *Quarterly Review of Biology*, *28*, 337–363.

Lewontin, R. C. (1982). Organism and environment. In H. C. Plotkin (Ed.), *Learning, development and culture: Essays in evolutionary epistemology*. (pp. 151–170). New York: Wiley.

Lumsden, C. J., & Wilson E. O., (1981). *Genes, mind and culture: The coevolutionary process*. Cambridge, MA: Harvard University Press.

Maier, N. R. F., & Schneirla, T. C., (1935). *Principles of animal psychology*. New York: McGraw-Hill.

Needham, J. (1929). *The skeptical biologist*. London: Chatto.

Novikoff, A. B., (1945). The concept of integrative levels and biology. *Science, 101*, 209–215.

Popper, K. (1974). Campbell on the evolutionary theory of knowledge. In *The library of living philosophers*. (Vol. XIV, Book II pp. 1059–1065.) LaSalle, Il: Open Court.

Prigogine, I. (1980). *From being to becoming: Time and complexity in the physical sciences*. San Francisco: W. H. Freeman.

Schneirla, T. C. (1949). Levels in the psychological capacities of animals. In L. R. Aronson, E. Tobach, J. S. Rosenblatt, & D. S. Lehrman (Eds.), *Selected writings of T. C. Schneirla, 1972*. San Francisco: W. H. Freeman.

Schneirla, T. C. (1951). The "levels" concept in the study of social organisation in animals. In L. R. Aronson, E. Tobach, J. S. Rosenblatt, & D. S. Lehrman (Eds.), *Selected writings of T. C. Schneirla, 1972*. San Francisco: W. H. Freeman.

Schneirla, T. C. (1965). Aspects of stimulation and organisation in approach-withdrawal processes underlying verebrate behavioral development. In L. R. Aronson, E. Tobach, J. S. Rosenblatt, & D. S. Lehrman (Eds.), *Selected Writings of T. C. Schneirla*, 1972. San Francisco: W. H. Freeman.

Tobach, E. (1972). The meaning of the cryptanthroparion. In L. Ehrman, G. Omenn, & E. Caspari (Eds.), *Genetics, environment and behavior* (pp. 219–239). New York: Academic.

Tobach, E. (1981). Evolutionary aspects of the activity of the organism and its development. In R. M. Lerner & N. A. Busch-Rossnagle (Eds.), *Individuals as producers of their development*. Orlando, FL: Academic.

Tobach, E. (1987). Integrative levels in the comparative psychology of cognition, language and consciousness. In G. Greenberg & E. Tobach (Eds.), *Cognition, language, and consciousness: Integrative levels*. Hillsdale, NJ: Lawrence Erlbaum Associates.

Webster's New World Dictionary (1988). V. Neufeldt & D. B. Guralnik (Eds.), *Webster's New World*, New York.

Wilson, E. O. (1978). *On human nature*. Cambridge, MA: Harvard University Press.

Author Index

Subject Index

H

I

K

L

M

N

O